Henry G. C Smith

**Practical Arithmetic for Senior Classes**

Henry G. C Smith

**Practical Arithmetic for Senior Classes**

ISBN/EAN: 9783337310554

Printed in Europe, USA, Canada, Australia, Japan

Cover: Foto ©berggeist007 / pixelio.de

More available books at **www.hansebooks.com**

# PRACTICAL ARITHMETIC

FOR

# SENIOR CLASSES.

BY

# HENRY G. C. SMITH,

TEACHER OF ARITHMETIC AND MATHEMATICS,
GEORGE HERIOT'S HOSPITAL.

SIXTH EDITION.

EDINBURGH:

OLIVER AND BOYD, TWEEDDALE COURT.

LONDON: SIMPKIN, MARSHALL, AND CO.

1871.

Price 2s. bound.—Answers to Ditto 6d.

Tɪɪ‍s Manual, which is a Sequel to *Practical Arithmetic for Junior Classes*, is intended for the use of those who have mastered the Fundamental Rules in Simple and Compound Numbers. Considerable space, in accordance with the importance of the subject, has been devoted to the explanation of Fractions; and the other branches also have been illustrated with a view to practical instruction. The Exercises, which are copious and original, have been constructed to combine interest with utility. They are arranged in distinct Sections, and are accompanied with Illustrative Processes. As the work is essentially practical, the explanatory remarks in elucidation of the various processes are more of an applicate than an abstract character.

The Answers to the Exercises in this Manual are published in a separate form.

# CONTENTS.

# ARITHMETICAL TABLES.

## I. MONEY.

### MONEY OF ACCOUNT.

| | | |
|---|---|---|
| 4 farthings | = 1 penny | *d.* |
| 12 d. | = 1 shilling | *s.* |
| 20 s. | = 1 pound | *£.* |

*d.* for *denarius* : *s.* for *solidus* :
*£* for *libra.*

### DECIMAL DIVISION OF £1.

| | | |
|---|---|---|
| 10 mils, *m.* | = 1 cent | *c.* |
| 10 c. | = 1 florin | *fl.* |
| 10 fl. | = 1 pound | *£.* |

### COINS IN CIRCULATION.

GOLD. *Sovereign*, £1 ; *Half-sov.* 10s. 1869 sovereigns are coined out of 40 lb. troy of sterling gold.

SILVER. *Crown*, 5s.; *Hf.-cr.* 2s. 6d.; *Florin*, 2s.; *Shilling*, 1s.; *Sixpence*, 6d.; *Groat*, 4d.; *Threepence*, 3d. 66 shillings are coined out of 1 lb. troy of sterling silver.

BRONZE. *Penny*, 1d.; *Halfpenny*, ½d.; *Farthing*, ¼d. In 100 parts of the bronze metal for these coins, there are :—95, copper; 4, tin; and 1, zinc. To 1 lb. avoir. there are :—Of pennies, 48; of halfpennies, 80; of farthings, 160.

### OBSOLETE COINS.

Tester, 6d.; Dollar, 4/6; Noble, 6/8; Seven Shilling piece; Angel, 10/; Half-guinea, 10/6; Mark, 13/4; Pistole, 16/; Guinea, 21/; Carolus, 23/; Jacobus, 25/; Moidore, 27/; Joannes, 36/.

The denominations of *Scots Money* are *one-twelfth* of the value of the corresponding names in sterling : thus,

| | | | |
|---|---|---|---|
| £1 | Scots | = | 20d. sterling. |
| 1s. | " | = | 1d. " |
| Also, 1 merk. " | | = | 13¼d. " |

## II. WEIGHT.

The Act 5° Geo. IV. cap. 74, which established the IMPERIAL WEIGHTS AND MEASURES, came into operation on 1st Jan. 1826. By the Act 18° and 19° Vic. cap. 72, the Imperial Standard of Weight is the *Pound Avoirdupois*, deposited in the Exchequer at Westminster, and of which copies are placed in the Mint, the Royal Society of London, Greenwich Observatory, and the Palace at Westminster.

### AVOIRDUPOIS WEIGHT.

Avoir. Wt. is the general weight of commerce. 1 lb. avoir. = 7000 grains.

| | | |
|---|---|---|
| 16 drams, *dr.* | = 1 ounce | *oz.* |
| 16 oz. | = 1 pound | *lb.* |
| 28 lb. | = 1 quarter | *qr.* |
| 4 qr. or 112 lb. | = 1 hundredwt. | *cwt.* |
| 20 cwt. | = 1 ton | *T.* |
| | Also, | |
| 14 lb. | = 1 stone | |

In London, a *stone* of butcher-meat = 8 lb. In Liverpool, &c., 100 lb. = 1 cental.

### TROY WEIGHT.

Troy Wt. is used in weighing the precious metals and in philosophical experiments.

| | | |
|---|---|---|
| 24 grains, *gr.* | = 1 pennywt. | *dwt.* |
| 20 dwt. | = 1 ounce | *oz.* |
| 12 oz. or 5760 gr. | = 1 pound | *lb.* |

At the Mint, the ounce is divided into 1000ths.

The fineness of gold is estimated in *carats.* Pure gold is said to be 24 carats fine. *Sterling* gold, of which every 24 parts contain 2 of alloy, is 22 carats fine.

The fineness of silver is estimated in oz. and dwt. *Sterling* silver, of which 1 lb. contains 18 dwt. of alloy, is 11 oz. 2 dwt. fine.

151½ *Diamond carats* = 1 oz. troy, which is also = 600 *Pearl grains.*

---

### APOTHECARIES' WEIGHT.

| | | |
|---|---|---|
| 20 grains, *gr.* | = 1 scruple | ℈ |
| 3 ℈ | = 1 drachm | ʒ |
| 8 ʒ | = 1 ounce | ℥ |

In the above, the ℥ is the ounce Troy of 480 grains; but in the NEW SYSTEM of weights adopted by the General Medical Council in October 1862, the ℈ and ʒ have been abolished, and the ℥ is *the ounce Avoirdupois of 437½ grains.*

---

### III. LENGTH.

By the Act 18° and 19° Vic. cap. 72, the Imperial Standard Measure of Length is the *Yard*, deposited in the Exchequer at Westminster, and of which copies are placed beside those of the Standard of Weight.

---

#### LINEAL MEASURE.

| | | |
|---|---|---|
| 12 lines, *l.* | = 1 inch | *in.* |
| 12 in. | = 1 foot | *ft.* |
| 3 ft. | = 1 yard | *yd.* |
| 5½ yd. | = 1 pole | *po.* |
| 40 po. | = 1 furlong | *fu.* |
| 8 fu. or 1760 yd. | = 1 mile | *ml.* |

Also

Length of 3 barleycorns = 1 in.
Breadth of 4 barleycorns = 1 digit = ¾ in.

| | | | | |
|---|---|---|---|---|
| Palm | = 3 | Cubit | = 1½ | ft. |
| Hand | = 4 | Step | = 2½ | |
| Span | = 9 | Pace | = 5 | |

Fathom = 6 ft.

---

#### GEOGRAPHICAL MEASURE.

| | |
|---|---|
| 6076 ft. *nearly* | = 1 geog. ml. |
| 60 geog. ml. | = 1 degree of the Earth's circumf. |
| 21600 geog. ml. | = the Earth's circ. |

Also

3 geog. ml. = 1 league

---

#### SURVEYORS' MEASURE.

| | | |
|---|---|---|
| 7$\frac{92}{100}$ in. | = 1 link | *lk.* |
| 100 lk. or 66 ft. | = 1 chain | *ch.* |
| 80 ch. | = 1 mile | *ml.* |

---

#### CLOTH MEASURE.

| | | |
|---|---|---|
| 2¼ in. | = 1 nail | *nl.* |
| 4 nl. or 9 in. | = 1 quarter | *qr.* |
| 4 qr. | = 1 yard | *yd.* |

Also

| | | |
|---|---|---|
| Flemish ell = 3 qr. | English ell = 5 qr |
| Scotch ˶ = 37 in. | French ˶ = 6 qr. |

---

### IV. SURFACE.

---

#### SQUARE MEASURE.

This Table is formed by *squaring* the corresponding denominations in Lineal Measure.

| | | |
|---|---|---|
| 144 sq. in. | = 1 sq. foot | *sq. ft.* |
| 9 sq. ft. | = 1 sq. yard | *sq. yd.* |
| 30¼ sq. yd. | = 1 sq. pole | *sq. po* |
| 40 sq. po. | = 1 rood | *ro.* |
| 4 ro. or 4840 s. yd. | = 1 acre | *ac.* |
| 640 ac. | = 1 sq. mile | *sq. ml.* |

Also

| | |
|---|---|
| 100 sq. ft. | = 1 square of flooring |
| 272¼ sq. ft. or 1 sq. po. } | = 1 rood of brickwork |
| 36 sq. yd. | = 1 rood of building |

---

#### SURVEYORS' MEASURE.

| | |
|---|---|
| 10,000 sq. lk. | = 1 sq. chain |
| 10 sq. ch. | = 1 acre |

---

### V. SOLIDITY.

---

#### CUBIC MEASURE.

This Table is formed by *cubing* the corresponding denominations in Lineal Measure.

| | |
|---|---|
| 1728 cub. in. | = 1 cub. ft. |
| 27 cub. ft. | = 1 cub. yd. |

Also

| | |
|---|---|
| 5 cub. ft. | = 1 barrel bulk *B. B.* |
| 8 B. B. | = 1 ton measurement |
| 40 cub. ft. of rough timber | = 1 load |
| 50 cub. ft. of hewn timber | = 1 load |
| 216 cub. ft. | = 1 cubic fathom |

## VI. CAPACITY.

According to the Act 5° Geo. IV. cap. 74, the Imperial Standard Measure of Capacity is the *Gallon*, which contains 10 lb. avoir. of distilled water weighed in air at the temperature of 62° Fahrenheit, the Barometer being at 30 in. The Standard Measure is deposited in the Exchequer at Westminster.

### MEASURE OF CAPACITY.

| | | |
|---|---|---|
| 4 gills *gi.* | = 1 pint | *pt.* |
| 2 pt. | = 1 quart | *qt.* |
| 4 qt. | = 1 gallon | *gal.* |
| 2 gal. | = 1 peck | *pk.* |
| 4 pk. | = 1 bushel | *bu.* |
| 8 bu. | = 1 quarter | *qr.* |

Also

| | | | |
|---|---|---|---|
| Pottle | = 2 qt | Coomb = | 4 bu. |
| Strike | = 2 bu. | Load = | 5 qr. |
| | Last | = 10 qr. | |

The *Imperial Gallon* = 277·274 cub. in., is the highest measure for liquids.

The weight of an *Imperial Bushel* of wheat varies from 56 lb. to 64 lb.: by the Tithe Commutation Act of England it is taken at 60 lb.

The following* were abolished by the Act 5° Geo. IV. cap. 74.

| | Cub. in. | Imperial |
|---|---|---|
| Wine Gallon | =231= | ·8331109 Gallon |
| Ale Gallon | =282=1·0170445 | ′ ′ |
| Winchester } Bushel } | =2150·42= | ·9694472 Bu. |

*Heaped Measure*, used for coals, &c., was abolished by the Act 5° and 6° Guliel. IV. cap. 63, which enacted that after 1st Jan. 1836, "all Coals, Slack, Culm, and Cannel of every Description, shall be sold by Weight and not by Measure." The bushel was = 1 Winchester bushel + 1 quart = 2217·62 cub. in., but when *heaped* in the form of a cone = 2815·486 cub. in.

| | |
|---|---|
| 3 heaped bushels | = 1 sack |
| 12 sacks | = 1 chaldron |

When the terms *Hogshead, Pipe*, &c. are used, it is merely as the names of

---

* The Weights and Measures of the United States of America are the same as those used in Great Britain, with the exception of the Measures of Capacity, which continue to be the subdivisions and multiples of the Winchester Bushel for dry goods, and of the Wine Gallon for liquids.

casks of *wine*, &c., and not as measures, for the contents must always be expressed in *Imperial Gallons*. When the names *Puncheon, Tierce*, are applied to casks of sugar, molasses, &c., their gross and net weight* must be stated.

---

### APOTHECARIES' FLUID MEASURE.

| | |
|---|---|
| 60 minims ℳ | =1 fluid drachm f ℨ |
| 8 f ℨ | =1 fluid ounce f ℥ |
| 20 f ℥ | =1 pint O |
| 8 O | =1 gallon C |

O for *Octarius ;* C for *Congius.*

1 f ℥ of distilled water weighs 1 oz. avoir.

---

## VII. INCLINATION.

### ANGULAR MEASURE.

| | | |
|---|---|---|
| 60 seconds ″ | = 1 minute | ′ |
| 60′ | = 1 degree | ° |
| 90° | = 1 right angle | L |
| 4 L or 360° | = 1 circle | ⊙ |

Also

| | |
|---|---|
| 30° | = 1 sign of the zodiac |

---

## VIII. TIME.

### MEASURE OF TIME.

| | | |
|---|---|---|
| 60 seconds, *sec.* | = 1 minute | *min.* |
| 60 min. | = 1 hour | *ho.* |
| 24 ho. | = 1 day | *da.* |
| 7 da. | = 1 week | *wk.* |
| 4 wk. | = 1 common month | *co.mo.* |
| 365 da. ⎱ or 52 w. ⎰ 1 d. | = 1 common year | *co. ye.* |
| 365 da. 6 ho. | = 1 Julian year | *Ju. yr.* |
| 366 da. | = 1 leap year | |

The year is divided into 12 *calendar* months :—

| | Da. | | Da. |
|---|---|---|---|
| January | 31 | July | 31 |
| February | 28 | August | 31 |
| March | 31 | September | 30 |
| April | 30 | October | 31 |
| May | 31 | November | 30 |
| June | 30 | December | 31 |

In leap year, February has 29 days.

---

* See *Allowances on Goods.*

A 2

QUARTERLY TERMS IN ENGLAND.

| | |
|---|---|
| Lady Day | March 25 |
| Midsummer | June 24 |
| Michaelmas Day | Sep. 29 |
| Christmas | Dec. 25 |

*Easter Day*, on which the Movable Feasts depend, is the first Sunday after the Paschal Full Moon, which happens on March 21, or next after it. When the Full Moon is on a Sunday, Easter Day is on the next Sunday.

QUARTERLY TERMS IN SCOTLAND.

| | |
|---|---|
| Candlemas | Feb. 2 |
| Whitsunday | May 15 |
| Lammas | Aug. 1 |
| Martinmas | Nov. 11 |

*The Sidereal Day* is = 23 ho. 56 min. 4·09 sec. It is the true time of the earth's revolution on its axis, or the interval between two successive meridian transits of the same star. A *sidereal* clock is always kept in an astronomical observatory.

*The Apparent Solar Day* is the interval between two successive meridian transits of the sun's centre. This day *varies* in length. The difference between *Apparent Solar Time* as shown by a sundial, and *Mean Solar Time* as indicated by a well-regulated clock, is termed Equation of Time.

*The Mean Solar Day* of 24 hours is used for the purposes of civil life. Astronomers in using the Mean Solar Day begin at 12 o'clock noon, and reckon the hours onward to 24. The *Astronomical* agrees with the *Civil Reckoning* from noon to midnight; but from midnight to noon the former is a day behind, thus:—

| Civil Time. | Astron. Time. |
|---|---|
| Sep. 10 : 7 p.m. = | Sep. 10 : 7 ho. |
| Sep. 11 : 11 a.m. = | Sep. 10 : 23 ho. |

Since the sun apparently describes a circle or 360° in 24 hours, 15° of longitude correspond to 1 hour of mean solar time; thus the time at a place in 45° E. long. is 3 hours before that of Greenwich, while in 60° W. long. it is 4 hours behind it.

*The Periodical Month* or sidereal revolution of the moon is = 27 da. 7 ho. 43 min. 11·5 sec. It is the time of the moon's revolution round the earth, or the interval in which the moon returns to the same place in the heavens.

*The Lunar Month* or synodical revolution of the moon is = 29 da. 12 ho. 44 min. 2·87 sec. It is the interval between new moon and new moon, or between two successive conjunctions of the sun and moon.

The Jews use a year of 12 lunar months of 29 or 30 days each; and to make it somewhat correspondent to the solar year, intercalate a month of 29 days, 7 times in a cycle of 19 years. The Mohammedans use a year of 12 lunar months or 354 days, and add a day to the year 11 times in 30 years.

*The Sidereal Year* is = 365 da. 6 ho. 9 min. 9·6 sec. It is the time of the earth's revolution round the sun.

*The Solar or Tropical Year* is = 365 da. 5 ho. 48 min. 49·7 sec. = 365·24224 days. It is the interval between two successive passages of the sun through the vernal equinox. The solar year regulates the seasons, and is therefore the proper standard for regulating the civil year.

Julius Cæsar adopted a nominal year of 365 da. 6 ho. In the Julian Calendar, every year whose number is divisible by 4 contains 366 days. The Julian Calendar was introduced 45 B.C. Its error is = 365·25 da. — 365·24224 da. = ·00776 da. ☿ yr., or 3·104 da. in 400 years. In the 16th century an error of 12 days had accumulated, but as it was determined to reckon merely from 325 A.D.— the year of the Council of Nice—Gregory XIII. ordered *ten* days to be omitted in October 1582. In the Gregorian Calendar, every year whose number is divisible by 4 is a leap year, except when divisible by 100 and not by 400; thus, while 1600 is a leap year, 1700, 1800, and 1900, are common years. 400 years in the Gregorian Calendar or New Style (N. S.) are thus 3 days shorter than 400 years in the Julian Calendar or Old Style (O. S.). The error of the Gregorian Calendar in 400 years is therefore 3·104 da. — 3 da. = ·104 da., or ·00026 da. ☿ yr. N. S. was introduced into the British Empire in September 1752. O. S. is still used by the Greek Church. The difference between O. S. and N. S. is *progressive*. In the 16th and 17th centuries it was *ten* days; in the 18th, *eleven;* in the 19th, *twelve.*

## MEMORANDA.

| | | | |
|---|---|---|---|
| Sack of Flour or Meal | | = | 280 lb. |
| Barrel | ″ | = | 196 ″ |
| Quire of Paper | | = | 24 sheets |
| Ream | ″ | = | 20 quires |
| Bale | ″ | = | 10 reams |
| Roll of Parchment | | = | 60 skins |
| Pack of Wool | | = | 240 lb. |
| Long hundred | | = | 120 |
| Gross | | = | 144 |

## METRIC SYSTEM OF WEIGHTS AND MEASURES.

The Use of the Metric System was rendered permissive in the United Kingdom by the Act 27° and 28° Vic. cap. 117.

When the metre was first definitively introduced in France in 1799, it was adopted as the ten-millionth part of the Quadrant from the N. Pole to the Equator, but subsequent calculations have, however, shown that it is not precisely so.

### MEASURES OF LENGTH.

| | Metre. | Inches. | | | Metres. | Yards. |
|---|---|---|---|---|---|---|
| Millimetre = | ·001 = | ·03937079 | | Dekametre = | 10 = | 10·93633 |
| Centimetre = | ·01 = | ·3937079 | | Hectometre = | 100 = | 109·3633 |
| Decimetre = | ·1 | = 3·937079 | | Kilometre = | 1000 = | 1093·633 |
| METRE = | 1· | = 39·37079 | | Myriametre = | 10000 = | 10936·33 |

### MEASURES OF SURFACE.

| | Sq. Metres. | Sq. Yards. | | | Sq. Metres. | Acres. |
|---|---|---|---|---|---|---|
| Centiare = | 1 = | 1·19603326 | | Dekare = | 1000 = | ·2471143 |
| ARE = | 100 = | 119·603326 | | Hectare = | 10000 = | 2·471143 |

### MEASURES OF CAPACITY.

| | Cub. Metre. | Pint. | | | Cub. Metre. | Gallons. |
|---|---|---|---|---|---|---|
| Centilitre = | ·00001 = | ·0176077 | | Dekalitre = | ·01 = | 2·20097 |
| Decilitre = | ·0001 = | ·176077 | | Hectolitre = | ·1 = | 22·0097 |
| LITRE = | ·001 = | 1·76077 | | Kilolitre = | 1· = | 220·097 |

### WEIGHTS.

| | Gram. | Grains. | | | Grams. | Pounds Avoir. |
|---|---|---|---|---|---|---|
| Milligram = | ·001 = | ·0154323487 | | Dekagram = | 10 = | ·022046212 |
| Centigram = | ·01 = | ·154323487 | | Hectogram = | 100 = | ·22046212 |
| Decigram = | ·1 = | 1·54323487 | | Kilogram = | 1000 = | 2·2046212 |
| GRAM = | 1· | = 15·4323487 | | Myriagram = | 10000 = | 22·046212 |
| | | | | Quintal = | 100000 = | 220·46212 |
| | | | | Millier = | 1000000 = | 2204·6212 |

## SCOTCH WEIGHTS AND MEASURES.

These were declared obsolete by the Act 5° Geo. IV. cap. 74.

### WEIGHT.

| | |
|---|---|
| 16 drops | = 1 ounce |
| 16 ounces | = 1 pound |
| 16 pounds | = 1 stone |

There were two kinds of weight:— *Troyes or Dutch Weight*, of which 1 lb. = 7608·95 Imperial grains, and *Tron Weight*, of which 1 lb. = 9622·67 Imperial grains. The *Standard Stone Troyes* was assigned to *Lanark*.

### LINEAL MEASURE.

| | |
|---|---|
| 37 inches = 1 ell | 4 falls = 1 chain |
| 6 ells = 1 fall | 80 chains = 1 mile |

The *Standard Ell*, kept at *Edinburgh*, = 37·0598 Imp. in. The chain = 1·123024 Imp. chain = 74·1196 Imp. ft.

### SQUARE MEASURE.

| | |
|---|---|
| 36 sq. ells | = 1 sq. fall |
| 40 sq. falls | = 1 rood |
| 4 roods = 1 acre | = 1·261183 Imp. acre |

### LIQUID MEASURE.

| | |
|---|---|
| 4 gills = 1 mutchkin | 2 chopins = 1 pint |
| 2 mutchk. = 1 chopin | 8 pints = 1 gallon |

The *Standard Pint*, kept at *Stirling* = 104·2034 cub. in. = ·375814 Imp. gallon.

### DRY MEASURE.

| | |
|---|---|
| 4 lippies = 1 peck | 4 firlots = 1 boll |
| 4 pecks = 1 firlot | 16 bolls = 1 chalder |

There were two kinds of Dry Measure, the one for *wheat* and the other for *barley*, *oats*, &c. The *Standard Firlots* were kept at *Linlithgow*.

| | Cub. in. | Imp. bu. |
|---|---|---|
| Wheat Firlot = | 2214·3235 = | ·908256 |
| Barley Firlot = | 3230·3072 = | 1·456279 |

There was great diversity in the measures used in the various counties.

The Standard Scotch *boll* of meal is usually reckoned at 140 lb. avoir.

# PRIME AND COMPOSITE NUMBERS.

## 1.

### PRIME NUMBERS.

A NUMBER which cannot be divided by any other without leaving a remainder is termed a *Prime Number* or *Prime*; thus, 1, 2, 3, 5, 7, 11, 13, are primes.

A number composed of two or more primes multiplied together is termed a *Composite Number*; thus, 6, 15, 35, are composite numbers.

(1) Find the primes in the following series :—

1, 2, 3, ~~4~~, 5, ~~6~~, 7, ~~8~~, ~~9~~, ~~10~~, 11, ~~12~~, 13, ~~14~~.

By eliding every *second* number after 2, we cancel all numbers ⌐ *2. By eliding every *third* after 3, we cancel those ⌐ 3. The numbers *not elided* are *prime*.

This process, commonly known as ERATOSTHENES'† SIEVE, may be abridged as in the following examples :—

(2) Find the primes to 50.

1, 2, 3, 5, 7, ~~9~~, 11, 13, ~~15~~, 17, 19, ~~21~~, 23, ~~25~~, ~~27~~, 29, 31, ~~33~~, ~~35~~, 37, ~~39~~, 41, 43, ~~45~~, 47, ~~49~~.

Since 2 is the *only even prime*, we omit all the other even numbers. In eliding the composite numbers containing any prime, we need not test any below the *square of the prime*; for since all the lower composite numbers containing the prime contain a lower prime also, they must have been previously elided; thus we begin to elide those ⌐ 3 from 9, and those ⌐ 5 from 25.

For the same reason we divide by no prime whose square is > (greater than) the highest number in the series; thus we finish by eliding 49, which is ⌐ 7.

(3) Find the primes from 100 to 150.

101, 103, ~~105~~, 107, 109, ~~111~~, 113, ~~115~~, ~~117~~, ~~119~~, ~~121~~, ~~123~~, ~~125~~, 127, ~~129~~, 131, ~~133~~, ~~135~~, 137, 139, ~~141~~, ~~143~~, ~~145~~, ~~147~~, 149.

---

* The sign ⌐, for " *divisible by*," was introduced by Mr Barlow of Woolwich Academy in 1811.

† Eratosthenes, curator of the Alexandrian Library, died B.C. 194.

**1.**  Elide every *third* number after 105, the first number ᴜⱶ 3
  „    „   *fifth*  „     „  105, „  „   „    „ 5.
  „    „   *seventh* „    „  105, „  „   „    „ 7.
  „    „   *eleventh* „   „  121, „  „   „    „ 11.
All the composites are now elided, for 149 is < (less than) the
square of the next prime, 13.

Find the primes below 1000, giving those in *each hundred*
as a separate exercise.

---

**2.**                PRIME FACTORS.

THE primes that make up a composite number are termed its
*Prime Factors;* thus, 2, 2, 2, 3, 7, are the prime factors of 168.

Resolve the following into prime factors.

$$2|168 = \begin{cases} 2\times2\times2\times3\times7 \\ \text{or } 2^3\times3\times7 \end{cases} \qquad 2|924 = \begin{cases} 2\times2\times3\times7\times11 \\ \text{or } 2^2\times3\times7\times11 \end{cases}$$

$$2\underline{|84} \qquad\qquad\qquad 2\underline{|462}$$
$$2\underline{|42} \qquad\qquad\qquad 3\underline{|231}$$
$$3\underline{|21} \qquad\qquad\qquad 7\underline{|77}$$
$$7 \qquad\qquad\qquad\qquad 11$$

| | | | | | |
|---|---|---|---|---|---|
| 1. 6 | 6. 42 | 11. 98 | 16. 143 | 21. 245 | 26. 624 |
| 2. 12 | 7. 55 | 12. 100 | 17. 154 | 22. 264 | 27. 1188 |
| 3. 15 | 8. 66 | 13. 105 | 18. 165 | 23. 275 | 28. 1331 |
| 4. 21 | 9. 70 | 14. 110 | 19. 192 | 24. 343 | 29. 1452 |
| 5. 30 | 10. 75 | 15. 125 | 20. 242 | 25. 539 | 30. 1584 |

Resolve the following into prime factors, and combine them
into sets of three factors each, not greater than 12.

$810=2\times3^4\times5$
$\quad=3^2\times3^2\times2\times5$   It is easier to obtain the   $9|810=9\times9\times10$
$\quad=9\times9\times10$   three factors thus :—   $\overline{90=9\times10}$

| | | | | | |
|---|---|---|---|---|---|
| 31. 225 | 35. 495 | 39. 405 | 43. 704 | 47. 240 | 51. 960 |
| 32. 315 | 36. 616 | 40. 448 | 44. 756 | 48. 486 | 52. 1152 |
| 33. 392 | 37. 968 | 41. 504 | 45. 792 | 49. 729 | 53. 1296 |
| 34. 441 | 38. 1089 | 42. 672 | 46. 1056 | 50. 768 | 54. 1728 |

---

**3.**          GREATEST COMMON MEASURE.

A NUMBER which divides another without leaving a remainder
is termed a *Measure* or *Factor* of that number; thus, 8 is a
measure of 24.  A number which divides two or more numbers without leaving a remainder is termed a *Common Measure*
of those numbers; thus, 6 is a common measure of 24 and 36.

**3.** The greatest number which divides two or more numbers without leaving a remainder is termed their *Greatest Common Measure* (G. C. M.); thus, 12 is the G. C. M. of 24 and 36; 6 the G. C. M. of 24, 36, and 54.

Numbers whose G. C. M. is 1 are *prime to each other*. Composite numbers may be prime to each other; thus, 25 is prime to 36.

(1) Find the G. C. M. of 78 and 300.

### I. By Prime Factors.

$$78 = 2 \times 3 \times 13; \quad 300 = 2^2 \times 3 \times 5^2$$
$$\text{G. C. M.} = 2 \times 3 = 6$$

Since 2 and 3 are the only factors *common* to 78 and 300, $2 \times 3$ or 6 is the G. C. M. of 78 and 300.

### II. By Division.

```
78)300(3        1 | 78 | 300 | 3
   234            | 66 | 234 |
   ‾‾‾          2 | 12 | 66  | 5
   66)78(1        | 12 | 60  |
      66          ‾‾‾‾‾‾‾‾‾‾‾‾‾
      ‾‾          |    |  6  | G. C. M.
      12)66(5
         60
G. C. M. ‾6)12(2
          12
```

6 is a common measure of 78 and 300. For, 6 measures $2 \times 6$ or 12; $5 \times 12$ or 60; $60 + 6$ or 66; $66 + 12$ or 78; $3 \times 78$ or 234; and $234 + 66$ or 300.

No number $> 6$ is a common measure of 78 and 300. Since every measure of 78 measures $3 \times 78$ or 234, every common measure of 78 and 300 measures 234 and 300. Now if a number is contained a certain number of times in 234 and another number of times in 300, the difference between the quotients is an integer, which is the number of times the number is contained exactly in $300 - 234$. Every common measure of 78 and 300 therefore measures $300 - 234$ or 66. Since it measures 78 and 66, it measures also $78 - 66$ or 12; hence also $5 \times 12$ or 60; and finally $66 - 60$ or 6. No common measure of 78 and 300 can therefore be $> 6$; but 6 is a common measure of 78 and 300; $\therefore$ (hence) 6 is their G. C. M.

### Find G. C. M. of

| | | | |
|---|---|---|---|
| 1. 48, 78 | 8. 841, 899 | 15. 1850, 1517 | 22. 3243,37976 |
| 2. 56, 98 | 9. 961, 1178 | 16. 1792, 1847 | 23. 31484,109268 |
| 3. 121, 143 | 10. 1243, 1469 | 17. 3927, 5049 | 24. 82739,57693 |
| 4. 342, 665 | 11. 1001, 1287 | 18. 1287, 1551 | 25. 10759,20405 |
| 5. 448, 784 | 12. 1131, 2639 | 19. 1537, 1802 | 26. 714285,857142 |
| 6. 203, 261 | 13. 9889, 986 | 20. 3056, 3629 | 27. 49593,43902 |
| 7. 375, 525 | 14. 1792, 1832 | 21. 1261, 22116 | 28. 17641,22243 |

(2) Find G. C. M. of 42, 56, and 49.

```
42)56(1              14)49(3
   42                   42
   14)42(3    G. C. M. 7)14(2
      42                14
```

Every C. M. of 42 and 56 is a measure of their G. C. M., 14; ∴.
every C. M. of 14 and 49 is a C. M. of 42, 56, and 49; but 7 is the
G. C. M. of 14 and 49; ∴. 7 is G. C. M. of 42, 56, and 49.

(3) Find G. C. M. of 192, 56, 44, 128, 94.

Take any two numbers, as 44 and 94; 2 is their G. C. M.  The
G. C. M. required cannot therefore be > 2.  Now 2 measures the
numbers 192, 56, 128; ∴. 2 is the G. C. M. required.

Suppose we had selected 56 and 128, their G. C. M. is not a
measure of all the rest.  G. C. M. of 8 and 44 is 4.  G. C. M. of 4
and 94 is 2, which measures 192, and 2 is the G. C. M. required.

To abridge the process, it is expedient to select at first two numbers whose G. C. M. is among the *least of the mutual* G. C. measures.

(4) Find G. C. M. of 27, 216, 48, 105, and 405.

3 is G. C. M. of 27 and 48; 3 is a M. of 216, 105, and 405.
∴. 3 is the G. C. M. of 27, 48, 216, 105, and 405.

Find G. C. M. of

| | | |
|---|---|---|
| 29. 45, 27, 54 | 32. 24, 36, 48, 216 | 35. 198, 495, 209, 660 |
| 30. 90, 84, 81 | 33. 32, 40, 64, 108 | 36. 146, 730, 365, 219 |
| 31. 56, 84, 63 | 34. 72, 84, 66, 176 | 37. 924, 378, 612, 246 |

38. Find the Greatest Common Divisor of 12460 and 10769.

39. Find the greatest number cancelling 1859 and 3003.

40. Find the length of the greatest line exactly measuring the
sides of an enclosure 216 yd. long and 111 broad.

41. Find the greatest measure of capacity contained exactly in
two measures containing respectively 6 gal. 7 pt. and 8 gal. 6 pt.

42. What is the greatest sum of money contained exactly in
£2 ″ 9 ″ 1 and £2 ″ 3 ″ 11?

43. Find the greatest sum of money contained exactly in
£34 ″ 7 ″ 7 and £70 ″ 12 ″ 2.

44. George, James, and John, wish to spend 2/6, 1/10½, and
3/5¼, on the *same* kind of squibs.  Find the price of the *dearest* squib
they can purchase.

45. Two apprentices carry 1147 and 961 ivory balls respectively
from the workshop to the showroom.  The balls are carried in
baskets of *equal* contents, which are filled and emptied several times.
How many balls are in a basketful?

**3.**   46. Two frigates having the *same* number of guns fire a number of *rounds*. The one has fired 608, and the other 1102 shots. How many guns has each ?

47. The *Nemesis* and *Mœander* frigates having the *same* number of guns greater than 36, fire a number of *rounds*. The one has fired 352 and the other 484 shots. How many guns has each ? And why is the limitation "*greater than* 36" necessary ?

48. Two opposition coaches, which have run *full* during the season for the *same* number of days, have had 4807 and 3971 passengers respectively. How many days has the season lasted, and how many passengers does the one contain more than the other ?

---

**4.**          LEAST COMMON MULTIPLE.

A NUMBER which contains another an exact number of times is termed a *Multiple* of that number; thus 48 is a Multiple of 8.

*Measure* and *Multiple* are correlative terms :—
    7 is a measure of 14, 14 is a multiple of 7.

A number containing two or more numbers an exact number of times is termed a *Common Multiple* of those numbers; thus 48 is a Common Multiple of 4, 6, and 8.

The least number containing two or more numbers an exact number of times is termed their *Least Common Multiple* (L. C. M.); thus 24 is L. C. M. of 4, 6, and 8.

When two or more numbers are prime to each other, their L. C. M. is their product; thus L. C. M. of 3, 5, 7, and 11, is $3 \times 5 \times 7 \times 11$.

(1) Find L. C. M. of 15 and 21.
$$15 = 3 \times 5; \; 21 = 3 \times 7.$$
$$\text{L. C. M. } 105 = 3 \times 7 \times 5.$$

Every common multiple of 15 and 21 must contain 3, 5, and 7. But $3 \times 5 \times 7$ is the least number containing 3, 5, and 7; ∴ $3 \times 5 \times 7$ is L. C. M. of 15 and 21.

L. C. M. of two numbers = Product ÷ G. C. M.

Thus, of 15 and 21; G. C. M. = 3; Product = $(3 \times 5) \times (3 \times 7)$.
$$\text{L. C. M.} = \frac{(3 \times 5) \times (3 \times 7)}{3} = 5 \times (3 \times 7).$$

In finding the L. C. M. of 2 numbers it is thus easier to divide one of the numbers by their G. C. M., and multiply the quotient by the other number.

(2) Find L. C. M. of 224 and 256.

$$\text{G. C. M.} = 32.$$
$$\text{L. C. M.} = \tfrac{224}{32} \times 256 = 7 \times 256 = 1792.$$

(3) Find L. C. M. of 384 and 564.

$$\text{G. C. M.} = 12.$$
$$\text{L. C. M.} = 32 \times 564 = 18048.$$

Find L. C. M. of

| | | | |
|---|---|---|---|
| 1. 27, 36 | 4. 72, 48 | 7. 144, 180 | 10. 200, 250 | 13. 420, 798 |
| 2. 42, 56 | 5. 52, 78 | 8. 216, 225 | 11. 224, 343 | 14. 225, 375 |
| 3. 35, 49 | 6. 34, 51 | 9. 196, 343 | 12. 324, 360 | 15. 234, 390 |

(4) Find L. C. M. of 16, 18, 21, 24, 30, 32, 36.

### I.

| 2 | ~~16,~~ | ~~18,~~ | 21, | 24, | 30, | 32, | 36 |
|---|---|---|---|---|---|---|---|
| 2 | | | 21, | 12, | 15, | 16, | 18 |
| 2 | | | 21, | 6, | 15, | 8, | 9 |
| 3 | | | 21, | ~~3,~~ | 15, | 4, | 9 |
| | | | 7, | | 5, | 4, | 3 |

We elide 16 and 18, which are respectively measures of 32 and 36. We divide by the prime 2 so long as it is contained in more than one number. Since 2 is not contained in 21, we continue to write 21 until we divide by the next prime. In the 4th line we elide 3, a measure of 9.

The factors contained in the numbers in addition to 2, 2, 2, 3, are 7, 5, 4, 3; and as these are prime to each other, L. C. M. = $2 \times 2 \times 2 \times 3 \times 7 \times 5 \times 4 \times 3 = 10080.$

### II.

| 12 | ~~16,~~ | ~~18,~~ | 21, | 24, | 30, | 32, | 36 |
|---|---|---|---|---|---|---|---|
| | | | 7, | ~~2,~~ | 5, | 8, | 3 |

Since the factors of 12 are contained in one or other of the numbers, we may divide by 12, and find the other factors contained in the numbers.

12 is not a measure of 21, but on dividing 21 by 3, the G. C. M. of 12 and 21, we obtain 7. Similarly we divide 30 and 32 respectively by 6 and 4.

The factors contained in the numbers besides those of 12, are 7, 5, 8, and 3; and as these are prime to each other, L. C. M. = $12 \times 7 \times 5 \times 8 \times 3 = 10080.$

In the First Method we divide by a prime so long as it is contained in two or more numbers. In the Second, we divide by a composite number whose factors are contained in one or other of the numbers; and when any number is not a multiple of the divisor, we divide it by their G. C. M.

**4.**     (5) Find L. C. M. of 21, 24, 25, 27, 28.

I.

$$2 \mid 21, \quad 24, \quad 25, \quad 27, \quad 28$$
$$2 \mid \overline{21, \quad 12, \quad 25, \quad 27, \quad 14}$$
$$3 \mid \overline{21, \quad 6, \quad 25, \quad 27, \quad 7}$$
$$\overline{7, \quad 2, \quad 25, \quad 9}$$

L. C. M. $= 2 \times 2 \times 3 \times 7 \times 2 \times$
$25 \times 9 = 37800.$

II.

$$12 \mid 21, \quad 24, \quad 25, \quad 27, \quad 28$$
$$\overline{7, \quad 2, \quad 25, \quad 9, \quad 7}$$

L. C. M. $= 12 \times 7 \times 2 \times 25 \times 9$
$= 37800.$

16. 4, 6, 10, 12
17. 8, 12, 15, 18
18. 12, 16, 18, 20
19. 12, 16, 18, 27
20. 10, 6, 15, 12
21. 12, 15, 20, 40
22. 12, 28, 35, 21

23. 32, 36, 49, 56, 42
24. 20, 24, 25, 27, 45
25. 28, 30, 32, 36, 42
26. 35, 40, 42, 49, 28
27. 8, 14, 18, 21, 32, 28
28. 24, 27, 28, 32, 36, 56
29. 15, 21, 24, 27, 28, 35

30. 25, 32, 63, 40, 35, 56, 80
31. 30, 36, 32, 48, 40, 54, 63
32. 30.33, 36, 42, 48, 63, 55
33. 27, 36, 45, 54, 63, 72, 81
34. 35, 45, 56, 63, 40, 72, 28
35. 15, 21, 33, 24, 35, 40, 77
36. 56, 40, 24, 88, 55, 21, 33

37. Find the least number containing the 9 digits.

38. Find the shortest distance that three rods of 8 ft. 3 yd. and 4 yd. will exactly measure.

39. Find the content of the smallest vessel that may be exactly filled by using a gallon, a 10 pint, or a 12 pint measure.

40. A rides at 10 miles an hour, B drives at 6 miles an hour, and C walks at 3 miles an hour. Find the shortest distance they may all traverse in an exact number of hours.

41. Tom, Dick, and Jack, agree to spend the same sum in purchasing fire-wheels at the rate of 1½d., 4d., and 2¼d. respectively. What is the smallest sum they can expend?

42. What are the prime factors of L. C. M. of 12, 35, 28, and of 21, 15, 20?

43. Mention the prime factors of L. C. M. of 12, 28, 35, 21, 55, and of 15, 33, 20, 77, 44.

(6) Find the least number which, when separately divided by 2, 3, 4, 5, 6, always leaves the remainder 1.
Least Number ÷ 2, 3, 4, 5, 6, is L. C. M. of 2, 3, 4, 5, 6 = 60; 60 + 1 or 61 is the number required.

44. Find the least number which, when separately divided by 2, 3......7, leaves the remainder 1.

(7) Find the least number which, when separately divided by 2, 3......8, leaves the remainders 1, 2......7 respectively.
L. C. M. = 840; 840 − 1 or 839 is the number required.

45. Find the least number which, when separately divided by 2, 3......9, leaves the remainders 1, 2......8 respectively.

# VULGAR FRACTIONS.

If a unit is divided into *three* equal parts, and *two* of them are taken, the parts thus taken form *two-thirds of one unit.*

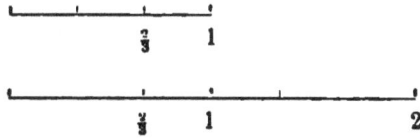

If *two* units as a whole are divided into *three* equal parts, and *one* taken, the part thus taken is *one-third of two units.*

That which we have obtained by either method is written, $\frac{2}{3}$. It is termed a VULGAR FRACTION, of which 2 is the *Numerator* and 3 the *Denominator.*

The Denominator of a Vulgar Fraction indicates the number of equal parts into which a unit is divided; and the Numerator, the number taken. Or, the Numerator indicates the number of units, and the Denominator the number of equal parts into which these units, considered as a whole, are divided, and of which one is to be taken.

**5.** If 2 units are each divided into 5 equal parts, we obtain 10 fifths. The integer 2 is thus reduced to the fractional form, $\frac{10}{5}$.

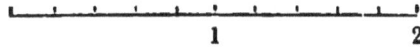

(1) Reduce 3 to an equivalent fraction with denominator 7.
$$3 = \frac{21}{7}.$$
1. Reduce 9 to equivalent fractions with denominators 4, 8, 7.
2.    „   33    „     „     „     „     3, 5, 8.
3.    „   29    „     „     „     „    11, 13, 20.
4.    „   37    „     „     „     „    12, 14, 15.

5. A baker divides 12 rolls into 4 equal parts each. How many fourths has he?

6. Into how many eighths of a yard can a draper cut 17 yards of cloth?

**6.** Suppose we have two units and three-fifths of a unit, by dividing each of the units into fifths, and adding in the three-fifths, we obtain thirteen-fifths.
$$2\tfrac{3}{5} = \frac{10+3}{5} = \frac{13}{5}.$$
Every number which, like $2\tfrac{3}{5}$, is thus made up of an integer and a fraction, is termed a MIXED NUMBER.

**6.** Reduce the following mixed numbers to a fractional form :—

$$79\tfrac{11}{13} = \frac{79 \times 13 + 11}{13} = \frac{1038}{13}.$$

| | | | | |
|---|---|---|---|---|
| 1. $7\tfrac{3}{4}$ | 4. $8\tfrac{6}{11}$ | 7. $90\tfrac{17}{50}$ | 10. $329\tfrac{13}{15}$ | 13. $928\tfrac{11}{13}$ |
| 2. $11\tfrac{4}{5}$ | 5. $8\tfrac{4}{13}$ | 8. $79\tfrac{33}{38}$ | 11. $924\tfrac{17}{17}$ | 14. $3\tfrac{16}{113}$ |
| 3. $13\tfrac{6}{7}$ | 6. $15\tfrac{16}{17}$ | 9. $23\tfrac{11}{38}$ | 12. $89\tfrac{33}{34}$ | 15. $23\tfrac{111}{331}$ |

16. How many eighths of a yard are in $7\tfrac{3}{8}$ yd. ?
17. How many twelfths of a penny are in $8\tfrac{6}{12}$d. ?
18. How many sixteenths of a yard has a draper sold who has disposed of $9\tfrac{3}{16}$ yd. ?

**7.** If we take the fraction $\tfrac{11}{4}$, we find we can make up two units, with three-fourths over.

$$\tfrac{11}{4} = 2\tfrac{3}{4}.$$

Every fraction which, like $\tfrac{11}{4}$, has its numerator $>$ its denominator, is $> 1$, and is termed an IMPROPER FRACTION.

Every fraction which, like $\tfrac{4}{4}$, has its numerator $=$ its denominator is $= 1$, and is termed an IMPROPER FRACTION.

Every fraction which, like $\tfrac{3}{4}$, has its numerator $<$ its denominator, is $< 1$, and is termed a PROPER FRACTION.

Reduce the following improper fractions to whole or mixed numbers :—

$$(1)\ \tfrac{168}{12} = 14. \quad (2)\ \tfrac{100}{13} = 7\tfrac{n}{13}. \quad (3)\ \tfrac{16}{16} = 1.$$

| | | | | |
|---|---|---|---|---|
| 1. $\tfrac{30}{9}$ | 4. $\tfrac{100}{31}$ | 7. $\tfrac{3191}{23}$ | 10. $\tfrac{1000}{27}$ | 13. $\tfrac{3217}{1250}$ |
| 2. $\tfrac{23}{13}$ | 5. $\tfrac{344}{113}$ | 8. $\tfrac{4100}{179}$ | 11. $\tfrac{4300}{41}$ | 14. $\tfrac{6000}{133}$ |
| 3. $\tfrac{149}{17}$ | 6. $\tfrac{2000}{43}$ | 9. $\tfrac{2983}{35}$ | 12. $\tfrac{1000}{73}$ | 15. $\tfrac{8000}{199}$ |

16. A grocer, who has sold 89 quarter-pounds of tea, wishes to know how many lb. he has sold.
17. A draper, who has sold 117 sixteenths of a yard, is asked how many yards he has sold.
18. The average length of a year, according to the Gregorian Calendar, is $\tfrac{146097}{400}$ days. Express this as a mixed number.

**8.** If we take any fraction, as $\tfrac{5}{7}$, all fractions with the denominator 7, having the numerator $< 5$, as $\tfrac{4}{7}$, $\tfrac{3}{7}$, &c., are $< \tfrac{5}{7}$; and all with the numerator 5, having the denominator $> 7$, as $\tfrac{5}{8}$, $\tfrac{5}{9}$, &c., are also $< \tfrac{5}{7}$. By diminishing

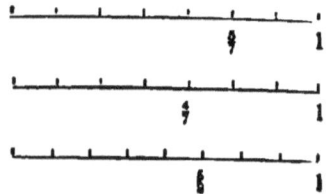

**8.** the numerator, or increasing the denominator, we thus *diminish* the value of a fraction.

Again, all fractions with the denominator 7, having the numerator $>$ 5, as $\frac{6}{7}$, $\frac{7}{7}$, &c., are $>$ $\frac{5}{7}$; and all with the numerator 5, but the denominator $<$ 7, as $\frac{5}{6}$, $\frac{5}{5}$, &c., are also $>$ $\frac{5}{7}$. By increasing the numerator, or diminishing the denominator, we thus *increase* the value of a fraction.

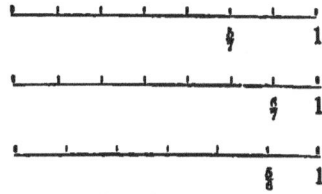

(1) Mention 4 fractions with denominator 12, next $>$ $\frac{5}{12}$.

$$\frac{6}{12}; \ \frac{6}{12}, \frac{7}{12}, \frac{8}{12}, \frac{9}{12}.$$

(2) Mention 3 fractions with numerator 5, next $<$ $\frac{5}{11}$.

$$\frac{5}{11}; \ \frac{5}{12}, \frac{5}{13}, \frac{5}{14}.$$

1. Mention 4 fractions with numerator   9, next $<$ $\frac{9}{11}$.
2.   "    3   "   "   "    6,  " $>$ $\frac{3}{17}$.
3.   "    6   "   " denominator 9,  " $>$ $\frac{6}{9}$.
4.   "    5   "   "   "    10,  " $<$ $\frac{7}{10}$.
5.   "    3   "   " numerator 13,  " $<$ $\frac{13}{3}$.
6.   "    3   "   " denominator 17,  " $<$ $\frac{17}{3}$.

**9.**   We may multiply $\frac{3}{7}$ by 2, by doubling the *number* of the parts; thus, $\frac{3}{7} \times 2 = \frac{6}{7}$.

We may multiply $\frac{3}{8}$ by 2, by doubling the *magnitude* of the parts; thus, $\frac{3}{8} \times 2 = \frac{3}{4}$.

By multiplying the numerator or dividing the denominator, we thus *multiply* the value of a fraction.

Multiply the following fractions by integers:—

$$(1) \ \frac{3}{12} \times 4 = \frac{12}{12}. \quad (2) \ \frac{3}{14} \times 7 = \frac{3}{2} = 1\frac{1}{2}. \quad (3) \ \frac{7}{60} \times 24 \overset{12 \times 2}{=} \frac{14}{5} = 2\frac{4}{5}.$$

In (3), 12, a common factor of 24 and 60, is contained 2 times in 24. To multiply $\frac{7}{60}$ by 24 we may therefore multiply the numerator by 2, and divide the denominator by 12.

| | | |
|---|---|---|
| 1. $\frac{1}{4} \times 5$ | 5. $\frac{3}{16} \times 3$ | 9. $\frac{3}{27} \times 12$ |
| 2. $\frac{2}{11} \times 4$ | 6. $\frac{13}{31} \times 4$ | 10. $\frac{3}{45} \times 14$ |
| 3. $\frac{3}{13} \times 4$ | 7. $\frac{11}{56} \times 7$ | 11. $\frac{12}{11} \times 27$ |
| 4. $\frac{2}{17} \times 8$ | 8. $\frac{13}{44} \times 16$ | 12. $\frac{44}{31} \times 15$ |

**9.**    13. Seven purchasers each buy $\frac{3}{4}$ peck of meal. How many pecks have been bought?

14. 1 lb. troy $= \frac{144}{175}$ lb. avoir. How many lb. avoir. $= 25$ lb. troy?

15. Find the number of degrees $= 25$ grades, of which each is $= \frac{9}{10}$ deg.

**10.**    We may divide $\frac{4}{5}$ by 4, by taking *one-fourth* of the *number* of parts; thus, $\frac{4}{5} \div 4 = \frac{1}{5}$.

We may divide $\frac{3}{5}$ by 2, by taking as many parts of *half the magnitude;* thus, $\frac{3}{5} \div 2 = \frac{3}{10}$.

By dividing the numerator or multiplying the denominator, we thus *divide* the value of a fraction.

Divide the following fractions by integers :—

(1) $\frac{3}{4} \div 5 = \frac{3}{20}$.    (2) $\frac{16}{23} \div 4 = \frac{4}{23}$.    (3) $\frac{15}{23} \div 18 = \frac{5}{138}$.
$\phantom{xxxxxxxxxxxxxxxxxxxxxxxxxxxxxxxxxxxxxxx}3 \times 6$

In (3), 3, a common factor of 18 and 15, is contained 6 times in 18. To divide $\frac{15}{23}$ by 18, we may therefore divide the numerator by 3, and multiply the denominator by 6.

| | | |
|---|---|---|
| 1. $\frac{4}{5} \div 2$ | 5. $\frac{14}{16} \div 5$ | 9. $\frac{24}{31} \div 20$ |
| 2. $\frac{9}{10} \div 3$ | 6. $\frac{22}{25} \div 3$ | 10. $\frac{32}{45} \div 12$ |
| 3. $\frac{12}{17} \div 6$ | 7. $\frac{14}{13} \div 8$ | 11. $\frac{14}{19} \div 21$ |
| 4. $\frac{12}{25} \div 9$ | 8. $\frac{13}{17} \div 6$ | 12. $\frac{18}{19} \div 30$ |

13. What part of a mile does a stream flow a minute which flows $\frac{4}{5}$ mile in 7 min.?

14. 12 oz. troy $= \frac{144}{175}$ lb. avoir. What part of 1 lb. avoir. is 1 oz. troy?

15. 64 squares of a draught-board occupy $\frac{12}{13}$ sq. ft. What does one square occupy?

**11.**    Having given any fraction, as $\frac{2}{3}$, by taking one-third of the number of parts each three times as large, we have the same fraction expressed, as $\frac{2}{3}$. By dividing the numerator by 3, we divide the fraction by 3; and by dividing the denominator by 3, we multiply the fraction by 3, and thus the fraction is unaltered in value.

By dividing the numerator and the denominator of a fraction by the same number, the value of the fraction remains *unaltered.*

The fraction $\frac{6}{9}$, when expressed as $\frac{2}{3}$, is said to be in its LOWEST TERMS. A fraction is in its lowest terms when its numerator and denominator are *prime to each other.*

**11.** Reduce the following fractions to their lowest terms :—

(1) $\frac{224}{784}$ $\begin{cases} \text{I. } \frac{224}{784} = \frac{84}{196} = \frac{12}{28} = \frac{3}{7}. \\ \text{II. } \frac{224}{784} = \frac{3}{7}. \end{cases}$

In I., we divide the numerator and denominator successively by the common factors 4, 7, 4, selected by *inspection*.

In II., we at once divide the numerator and denominator by their G. C. M., 112.

The product of the factors 4, 7, 4, used in I., is = G. C. M., 112.

(2) $\frac{224}{374}$.  By inspection, $\overset{2}{\overset{}{\frac{224}{374}}} = \overset{7}{\overset{}{\frac{112}{187}}} = \frac{16}{17}$.

(3) $\frac{1480}{3034}$.  $\overset{\text{G. C. M.}}{74}$ | $\frac{1480}{3034} = \frac{24}{41}$.

| | | | |
|---|---|---|---|
| 1. $\frac{24}{36}$ | 13. $\frac{2484}{4056}$ | 25. $\frac{3927}{50400}$ | 37. $\frac{2463}{30720}$ |
| 2. $\frac{48}{54}$ | 14. $\frac{1436}{1728}$ | 26. $\frac{15742}{51632}$ | 38. $\frac{6201}{14480}$ |
| 3. $\frac{96}{100}$ | 15. $\frac{812}{3045}$ | 27. $\frac{201}{25380}$ | 39. $\frac{777}{32190}$ |
| 4. $\frac{63}{81}$ | 16. $\frac{999}{5553}$ | 28. $\frac{5001}{18337}$ | 40. $\frac{13064}{44221}$ |
| 5. $\frac{72}{108}$ | 17. $\frac{1272}{4700}$ | 29. $\frac{841}{50000}$ | 41. $\frac{47619}{333333}$ |
| 6. $\frac{324}{414}$ | 18. $\frac{1911}{4700}$ | 30. $\frac{21384}{46083}$ | 42. $\frac{386714}{999999}$ |
| 7. $\frac{3630}{4455}$ | 19. $\frac{1792}{3584}$ | 31. $\frac{2599}{83584}$ | 43. $\frac{102564}{171171}$ |
| 8. $\frac{1221}{1875}$ | 20. $\frac{16201}{40733}$ | 32. $\frac{377}{70170}$ | 44. $\frac{305128}{333333}$ |
| 9. $\frac{1148}{1618}$ | 21. $\frac{811}{10007}$ | 33. $\frac{3021}{50350}$ | 45. $\frac{867142}{}$ |
| 10. $\frac{1001}{1287}$ | 22. $\frac{10212}{51738}$ | 34. $\frac{6448}{51345}$ | 46. $\frac{112341}{364646}$ |
| 11. $\frac{2424}{4747}$ | 23. $\frac{11880}{13088}$ | 35. $\frac{2484}{4554}$ | 47. $\frac{618384}{}$ |
| 12. $\frac{2311}{2811}$ | 24. $\frac{33277}{40610}$ | 36. $\frac{6339}{14791}$ | 48. $\frac{777777}{854700}$ |

**12.** Having given any fraction, as $\frac{4}{5}$, by taking twice as many parts of half the magnitude, we have the same fraction expressed, as $\frac{8}{10}$. By multiplying the numerator by 2, we multiply the fraction by 2; and by multiplying the denominator by 2, we divide the fraction by 2; and thus by multiplying the numerator and the denominator of a fraction by the same number, the value of the fraction remains *unaltered*.

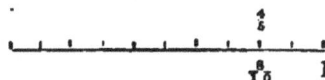

Take any two fractions, $\frac{3}{4}$, $\frac{5}{6}$. Of the Common Multiples of 4 and 6, let us take 24, $\frac{3}{4} = \frac{18}{24}$, and $\frac{5}{6} = \frac{20}{24}$. The fractions have thus been reduced to equivalent fractions with *a Common Denominator*.

12 is the L. C. M. of 4 and 6; $\frac{3}{4} = \frac{9}{12}$ and $\frac{5}{6} = \frac{10}{12}$. The fractions have thus been reduced to equivalent fractions with *the Least Common Denominator*.

**12.**  Reduce the following to equivalent fractions having the least common denominator (L. C. D.):—

(1) $\frac{2}{3}$, $\frac{5}{6}$, $\frac{7}{12}$, $\frac{15}{16}$.

L. C. D. = L. C. M. of 3, 6, 12, 16 = 48.

Since the denominator of $\frac{2}{3}$ is contained 16 times in 48, we multiply its numerator by 16. Similarly, we divide the L. C. D. by the denominator of each fraction, and multiply the corresponding numerator by the quotient. The number, showing how often the denominator of a fraction is contained in the L. C. D., is that by which the numerator of that fraction is multiplied, and may be termed the FRACTIONAL MULTIPLIER (F. M.)

F. M.

$16 \quad \dfrac{2}{3} = \dfrac{32}{48}$

$8 \quad \dfrac{5}{6} = \dfrac{40}{48}$

$4 \quad \dfrac{7}{12} = \dfrac{28}{48}$

$3 \quad \dfrac{15}{16} = \dfrac{45}{48}$

1. $\frac{1}{2}$, $\frac{2}{3}$, $\frac{3}{4}$

2. $\frac{2}{3}$, $\frac{3}{4}$, $\frac{5}{6}$

3. $\frac{2}{3}$, $\frac{3}{4}$, $\frac{5}{6}$, $\frac{7}{12}$

4. $\frac{1}{3}$, $\frac{1}{4}$, $\frac{5}{6}$, $\frac{5}{8}$

5. $\frac{4}{6}$, $\frac{8}{9}$, $\frac{11}{12}$, $\frac{13}{15}$

6. $\frac{3}{5}$, $\frac{7}{9}$, $\frac{7}{12}$, $\frac{11}{15}$

7. $\frac{5}{8}$, $\frac{8}{15}$, $\frac{13}{24}$, $\frac{19}{30}$

8. $\frac{7}{8}$, $\frac{17}{24}$, $\frac{19}{32}$, $\frac{11}{48}$

9. $\frac{2}{7}$, $\frac{3}{14}$, $\frac{5}{18}$, $\frac{3}{9}$, $\frac{2}{21}$

10. $\frac{3}{5}$, $\frac{5}{8}$, $\frac{9}{11}$, $\frac{3}{22}$, $\frac{6}{33}$

11. $\frac{3}{8}$, $\frac{3}{4}$, $\frac{3}{16}$, $\frac{3}{64}$, $\frac{3}{256}$

12. $\frac{7}{9}$, $\frac{7}{81}$, $\frac{7}{27}$, $\frac{7}{18}$, $\frac{7}{32}$

13. $\frac{1}{2}$, $\frac{3}{4}$, $\frac{7}{8}$, $\frac{15}{16}$, $\frac{31}{32}$, $\frac{63}{64}$

14. $\frac{7}{11}$, $\frac{7}{22}$, $\frac{7}{33}$, $\frac{7}{44}$, $\frac{7}{55}$, $\frac{7}{66}$

15. $\frac{3}{5}$, $\frac{7}{15}$, $\frac{2}{9}$, $\frac{11}{24}$, $\frac{7}{8}$, $\frac{17}{45}$

16. $\frac{13}{15}$, $\frac{13}{27}$, $\frac{13}{45}$, $\frac{13}{81}$, $\frac{13}{243}$, $\frac{13}{135}$

17. $\frac{2}{3}$, $\frac{3}{4}$, $\frac{5}{7}$, $\frac{7}{14}$, $\frac{13}{18}$, $\frac{4}{27}$

18. $\frac{11}{12}$, $\frac{11}{24}$, $\frac{11}{36}$, $\frac{11}{48}$, $\frac{11}{60}$, $\frac{11}{72}$

19. $\frac{4}{5}$, $\frac{5}{8}$, $\frac{11}{21}$, $\frac{14}{27}$, $\frac{23}{28}$, $\frac{37}{44}$

20. $\frac{4}{5}$, $\frac{24}{25}$, $\frac{124}{125}$, $\frac{624}{625}$, $\frac{3124}{3125}$

21. $\frac{3}{35}$, $\frac{7}{30}$, $\frac{11}{42}$, $\frac{9}{20}$, $\frac{7}{24}$, $\frac{2}{8}$

22. $\frac{5}{6}$, $\frac{5}{81}$, $\frac{5}{243}$, $\frac{7}{720}$, $\frac{5}{27}$, $\frac{5}{54}$

23. $\frac{11}{12}$, $\frac{13}{16}$, $\frac{17}{18}$, $\frac{35}{36}$, $\frac{31}{35}$, $\frac{23}{24}$

24. $\frac{7}{9}$, $\frac{8}{11}$, $\frac{5}{6}$, $\frac{3}{17}$, $\frac{280}{561}$, $\frac{7}{1122}$

(2) $\frac{6}{7}$, $\frac{9}{11}$, $\frac{4}{13}$.

L. C. D. = 7 × 11 × 13 = 1001.

When the denominators are prime to each other, the F. M. of a fraction is the product of the denominators of the other fractions: thus, the F. M. of $\frac{6}{7}$ is 11 × 13, which is = $\dfrac{7 \times 11 \times 13}{7}$.

F. M.

$11 \times 13 \quad \dfrac{6}{7} = \dfrac{858}{1001}$

$7 \times 13 \quad \dfrac{9}{11} = \dfrac{819}{1001}$

$7 \times 11 \quad \dfrac{4}{13} = \dfrac{308}{1001}$

(3) $\frac{3}{7}$, $\frac{1}{13}$, $\frac{7}{19}$, $\frac{12}{23}$.

L. C. D. = 7 × 13 × 19 × 23 = 39767.

Although the denominators are prime to each other, yet it is often convenient, when there are four or more fractions, to obtain the F. M. by dividing the L. C. D. by each denominator in succession: thus, the F. M. of $\frac{3}{7}$ is 5681 = $\dfrac{7 \times 13 \times 19 \times 23}{7}$, which is = 13 × 19 × 23.

F. M.

$5681 \quad \dfrac{3}{7} = \dfrac{17043}{39767}$

$3059 \quad \dfrac{1}{13} = \dfrac{3059}{39767}$

$2093 \quad \dfrac{7}{19} = \dfrac{14651}{39767}$

$1729 \quad \dfrac{12}{23} = \dfrac{20748}{39767}$

**12.**

| | |
|---|---|
| 25. $\frac{1}{8}, \frac{1}{4}, \frac{1}{5}$ | 31. $\frac{1}{2}, \frac{1}{3}, \frac{1}{5}, \frac{1}{7}, \frac{1}{11}$ |
| 26. $\frac{2}{3}, \frac{3}{4}, \frac{3}{7}$ | 32. $\frac{3}{4}, \frac{4}{7}, \frac{3}{5}, \frac{3}{11}$ |
| 27. $\frac{3}{4}, \frac{4}{5}, \frac{5}{7}, \frac{6}{11}$ | 33. $\frac{5}{8}, \frac{4}{7}, \frac{11}{25}, \frac{9}{13}$ |
| 28. $\frac{1}{3}, \frac{3}{5}, \frac{3}{7}, \frac{7}{11}$ | 34. $\frac{6}{7}, \frac{7}{8}, \frac{8}{9}, \frac{10}{11}$ |
| 29. $\frac{1}{2}, \frac{2}{3}, \frac{3}{11}, \frac{4}{13}$ | 35. $\frac{7}{8}, \frac{8}{9}, \frac{2}{5}, \frac{2}{13}$ |
| 30. $\frac{1}{3}, \frac{5}{8}, \frac{1}{11}, \frac{1}{17}$ | 36. $\frac{2}{3}, \frac{3}{4}, \frac{4}{7}, \frac{8}{11}, \frac{4}{5}$ |

**13.** In comparing the magnitudes of a number of fractions, as $\frac{5}{6}$, $\frac{7}{8}$, $\frac{7}{13}$, we may take a line, or, for the sake of distinctness, as many equal lines as there are fractions, and lay off parts corresponding to them. By an appeal to the eye, or by the aid of compasses, we may then compare the magnitude of the fractions.

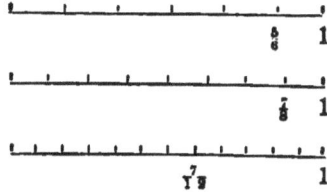

In comparing the fractions arithmetically, we may proceed as in the following examples.

Arrange the following fractions in order of magnitude :—

(1) $\frac{5}{6}$, $\frac{7}{8}$, $\frac{13}{14}$, respectively $= \frac{140}{144}, \frac{126}{144}, \frac{117}{144}$.

Order of Magnitude, $\frac{5}{6}$, $\frac{13}{14}$, $\frac{7}{8}$.

By reducing the fractions to a common denominator, we at once discover the order of magnitude.

(2) $\frac{3}{4}$, $\frac{5}{6}$, $\frac{6}{7}$, $\frac{9}{10}$, $\frac{7}{8}$.

Complements, $\frac{1}{4}$, $\frac{1}{6}$, $\frac{1}{7}$, $\frac{1}{10}$, $\frac{1}{8}$.

Order of Magnitude, $\frac{9}{10}$, $\frac{6}{7}$, $\frac{7}{8}$, $\frac{5}{6}$, $\frac{3}{4}$.

That which, when added to a proper fraction, makes up *unity*, is termed its COMPLEMENT. Of a number of fractions, that which has the least complement is the greatest fraction.

(3) $\frac{1}{13}$, $\frac{2}{27}$, $\frac{3}{40}$, respectively $= \frac{1}{13}, \frac{1}{13\frac{1}{2}}, \frac{1}{13\frac{1}{3}}$.

Order of Magnitude, $\frac{1}{13}$, $\frac{3}{40}$, $\frac{2}{27}$.

(4) $\frac{8}{9}$, $\frac{68}{76}$, $\frac{17}{18}$, $\frac{44}{45}$, respectively $= \frac{8}{9}, \frac{7\frac{5}{9}}{9}, \frac{8\frac{1}{2}}{9}, \frac{8\frac{4}{5}}{9}$.

Order of Magnitude, $\frac{44}{45}$, $\frac{17}{18}$, $\frac{8}{9}$, $\frac{68}{76}$.

Of the series of fractions $\frac{1}{13}$, $\frac{2}{20}$, $\frac{5}{34}$, $\frac{4}{35}$, arranged in order of magnitude, let any two, as $\frac{2}{20}$ and $\frac{5}{34}$, be taken. Reducing them to the common denominator $20 \times 34$, we have the numerators respectively $= 3 \times 34$ and $20 \times 5$, of which the former is the greater. When fractions are arranged in order of

B

**13.**    magnitude, the product of the numerator of any fraction by the denominator of the next less is $>$ the product of the denominator of the former fraction by the numerator of the latter; thus, in the series $1\frac{5}{6}$, $2\frac{3}{4}$, $4\frac{1}{6}$, $\frac{4}{5}$; $15 \times 26 > 16 \times 23$, $23 \times 36 > 26 \times 31$, $31 \times 6 > 36 \times 5$.

1. $\frac{1}{2}, \frac{2}{3}, \frac{3}{4}$

2. $\frac{1}{6}, \frac{2}{13}, \frac{1}{7}$

3. $\frac{4}{5}, \frac{24}{25}, \frac{5}{6}$

4. $\frac{3}{5}, \frac{11}{15}, \frac{7}{10}$

5. $\frac{3}{32}, \frac{1}{7}, \frac{2}{15}$

6. $\frac{2}{11}, \frac{5}{33}, \frac{13}{66}$

7. $\frac{11}{14}, \frac{9}{10}, \frac{14}{15}, \frac{13}{14}$

8. $\frac{2}{3}, \frac{11}{15}, \frac{7}{12}, \frac{5}{9}, \frac{11}{21}$

9. $\frac{3}{4}, \frac{4}{5}, \frac{7}{10}, \frac{7}{15}, \frac{7}{17}$

10. $\frac{4}{5}, \frac{11}{15}, \frac{17}{20}, \frac{49}{60}, \frac{5}{6}$

11. $\frac{15}{19}, \frac{13}{17}, \frac{14}{15}, \frac{19}{20}$

12. $\frac{9}{10}, \frac{15}{16}, \frac{19}{20}, \frac{16}{17}, \frac{12}{13}$

13. $\frac{10}{21}, \frac{7}{12}, \frac{3}{7}, \frac{4}{7}, \frac{4}{13}, \frac{4}{14}$

14. $\frac{13}{50}, \frac{3}{5}, \frac{4}{9}, \frac{5}{9}, \frac{19}{25}, \frac{9}{50}$

15. $\frac{39}{75}, \frac{79}{157}, \frac{17}{55}, \frac{42}{55}, \frac{8}{19}, \frac{7}{17}$

16. $\frac{2}{11}, \frac{7}{35}, \frac{6}{31}, \frac{9}{50}, \frac{11}{60}, \frac{3}{17}$

17. $\frac{3}{19}, \frac{4}{25}, \frac{5}{26}, \frac{2}{11}, \frac{16}{81}, \frac{17}{86}$

18. $\frac{1}{17}, \frac{6}{85}, \frac{6}{68}, \frac{3}{34}, \frac{4}{51}, \frac{2}{33}$

**14.**    Let us ADD the fractions $\frac{1}{2}$, $\frac{3}{4}$, and $\frac{5}{6}$. By taking any line as the unit, we place the lines representing $\frac{1}{2}$, $\frac{3}{4}$, $\frac{5}{6}$, in a line, and thus obtain their sum. To express the value of the sum, or to add the fractions arithmetically, we reduce them to equivalent fractions having a common denominator, and add the numerators of the equivalents. Of common denominators, the L. C. D. is generally taken.

(1) $\frac{1}{2} + \frac{3}{4} + \frac{5}{6}$.

$\frac{1}{2} = \frac{6}{12}$
$\frac{3}{4} = \frac{9}{12}$
$\frac{5}{6} = \frac{10}{12}$
$\frac{25}{12} = 2\frac{1}{12}$

*Otherwise*

$\frac{1}{2} + \frac{3}{4} + \frac{5}{6} = \frac{6+9+10}{12} = \frac{25}{12} = 2\frac{1}{12}$

(2) $\frac{1}{2} + \frac{7}{8} + \frac{11}{12} + \frac{4}{5} + \frac{8}{15} = \frac{60 + 105 + 110 + 96 + 64}{120}$
$= \frac{435}{120} = \frac{29}{8} = 3\frac{5}{8}$.

1. $\frac{1}{2} + \frac{1}{3} + \frac{1}{4}$

2. $\frac{5}{6} + \frac{7}{8} + \frac{5}{9}$

3. $\frac{3}{4} + \frac{7}{8} + \frac{9}{10}$

4. $\frac{8}{11} + \frac{7}{9} + \frac{4}{5}$

5. $\frac{9}{7} + \frac{8}{9} + \frac{4}{53}$

6. $\frac{3}{8} + \frac{4}{9} + \frac{17}{56}$

7. $\frac{11}{12} + \frac{7}{9} + \frac{17}{42} + \frac{10}{21}$

8. $\frac{9}{10} + \frac{3}{5} + \frac{6}{15} + \frac{7}{30}$

9. $\frac{8}{7} + \frac{2}{49} + \frac{3}{14} + \frac{6}{21}$

10. $\frac{11}{12} + \frac{3}{4} + \frac{7}{48} + \frac{11}{72}$

11. $\frac{5}{8} + \frac{7}{9} + \frac{13}{48} + \frac{11}{144}$

12. $\frac{13}{14} + \frac{3}{49} + \frac{16}{35} + \frac{11}{8}$

13. $\frac{3}{4} + \frac{11}{17} + \frac{19}{54} + \frac{13}{108} + \frac{17}{81} + \frac{19}{243}$

14. $\frac{7}{11} + \frac{3}{22} + \frac{9}{66} + \frac{13}{44} + \frac{13}{132} + \frac{79}{88}$

15. $\frac{11}{15} + \frac{16}{45} + \frac{9}{75} + \frac{17}{150} + \frac{13}{135} + \frac{11}{270}$

16. $\frac{3}{7} + \frac{4}{21} + \frac{5}{14} + \frac{9}{28} + \frac{11}{56} + \frac{11}{63}$

**14.**

(3) $1\frac{2}{3} + 1\frac{1}{2} = 2 + \frac{4+3}{6} = 2 + \frac{7}{6} = 3\frac{1}{6}.$

In the diagram, having represented $1\frac{1}{2}$ and $1\frac{2}{3}$ by lines, we place the integers together, and then the fractions in the same line with them, and thus obtain the sum whose value is $2 + \frac{7}{6} = 3\frac{1}{6}.$

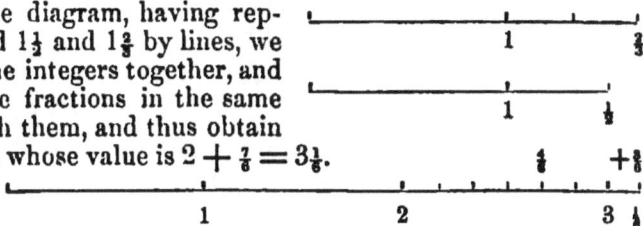

In adding Mixed Numbers we need not reduce them to improper fractions.

(4) $4\frac{1}{2} + 6\frac{11}{13} + 7\frac{8}{9} + 2\frac{11}{12} = 19 + \frac{234 + 396 + 416 + 429}{468}$

$= 19 + 3\frac{71}{468} = 22\frac{71}{468}.$

17. $6\frac{1}{4}+7\frac{3}{5}+8\frac{8}{9}+9\frac{5}{6}$

18. $3\frac{2}{3}+7\frac{1}{4}+8\frac{1}{4}+9\frac{7}{8}$

19. $7\frac{3}{5}+8\frac{3}{4}+5\frac{13}{16}+7\frac{11}{12}$

20. $5\frac{1}{2}+6\frac{3}{4}+7\frac{11}{24}+9\frac{17}{48}$

21. $9\frac{3}{4}+10\frac{5}{7}+11\frac{3}{8}+5\frac{17}{14}$

22. $11\frac{3}{8}+3\frac{1}{4}+2\frac{11}{25}+3\frac{4}{75}$

23. $\frac{17}{33}+\frac{14}{44}+\frac{19}{48}+7\frac{1}{4}+\frac{3}{18}+19\frac{23}{180}$

24. $3\frac{1}{14}+\frac{1}{4}+\frac{1}{21}+9\frac{4}{42}+2\frac{5}{81}+\frac{1}{63}$

25. $7\frac{1}{4}+8\frac{5}{6}+6\frac{1}{2}+7\frac{11}{16}+\frac{3}{32}+\frac{7}{64}$

26. $4\frac{11}{15}+\frac{3}{16}+5\frac{4}{55}+6\frac{7}{51}+\frac{1}{85}+\frac{1}{130}$

27. $\frac{11}{15}+\frac{7}{30}+\frac{4}{45}+6\frac{7}{90}+2\frac{7}{75}+\frac{11}{80}$

28. $7\frac{3}{8}+\frac{7}{4}+8\frac{1}{2}+11\frac{2}{3}+17\frac{3}{11}+\frac{1}{77}$

The work may be abridged by *combining*, in the process of adding, those fractions whose denominators are either the *same* or *have a common measure*, as in the following examples:

(5) $\frac{3}{10} + \frac{3}{18} + \frac{1}{10} + \frac{1}{4} + \frac{3}{5} + \frac{1}{18}.$

$\frac{3}{10} + \frac{1}{10} = \frac{4}{10} = \frac{2}{5}; \quad \frac{2}{5} + \frac{3}{5} = 1$

$\frac{3}{18} + \frac{1}{18} = \frac{4}{18} = \frac{2}{9}; \quad \frac{1}{4} + \frac{1}{4} = \underline{\frac{1}{2}}$

$\overline{1\frac{1}{2}}$

(6) $\frac{1}{8} + \frac{3}{17} + \frac{5}{24} + \frac{11}{51}.$

$\frac{1}{8} + \frac{5}{24} = \frac{3+5}{24} = \frac{8}{24} = \frac{1}{3}$

$\frac{3}{17} + \frac{11}{51} = \frac{6+11}{51} = \frac{17}{51} = \underline{\frac{1}{3}}$

$\overline{\frac{2}{3}}$

29. $\frac{1}{2}+\frac{3}{4}+\frac{3}{7}+\frac{3}{8}+\frac{4}{7}+\frac{1}{4}$

30. $\frac{1}{13}+\frac{3}{5}+\frac{1}{4}+\frac{2}{5}+\frac{1}{5}+\frac{3}{30}$

31. $\frac{4}{7}+\frac{3}{8}+\frac{1}{7}+\frac{3}{14}+\frac{7}{24}+\frac{3}{7}$

32. $\frac{1}{2}+\frac{1}{15}+\frac{3}{20}+\frac{1}{4}$

33. $1\frac{1}{4}+\frac{3}{7}+2\frac{9}{4}+\frac{9}{25}+\frac{5}{12}+\frac{1}{4}$

34. $\frac{3}{5}+\frac{3}{8}+\frac{15}{18}+\frac{17}{30}+\frac{7}{18}+\frac{5}{8}$

35. $\frac{17}{22}+\frac{3}{8}+\frac{3}{4}+\frac{5}{48}+\frac{9}{16}+\frac{1}{2}$

36. $\frac{5}{18}+\frac{3}{8}+\frac{5}{48}+\frac{1}{36}$

**14.**   (7) A student spends $\frac{1}{4}$ of the day in teaching, $\frac{1}{12}$ in attending classes, $\frac{5}{24}$ in study, $\frac{5}{48}$ in recreation and meals, and $\frac{1}{24}$ in miscellaneous reading. What part of the day is he thus occupied?

$$\tfrac{1}{4}+\tfrac{1}{12}+\tfrac{5}{24}+\tfrac{5}{48}+\tfrac{1}{24}=\frac{12+4+10+5+2}{48}$$
$$=\tfrac{33}{48}=\tfrac{11}{16}.$$

37. $\frac{1}{5}$ of a pole is in sand, and $\frac{4}{15}$ of it in water. What part of the pole is thus below the level of the water?

38. In an Allied Camp, $\frac{1}{5}$ of the soldiers are natives of England, $\frac{2}{15}$ of Scotland, $\frac{1}{190}$ of Ireland, and $\frac{1}{210}$ of Wales. What part of the camp is under British colours?

39. Of the chairs in the University of Edinburgh, $\frac{9}{34}$ of the number was founded in the nineteenth century, $\frac{1}{2}$ in the eighteenth, and $\frac{7}{34}$ in the seventeenth. What part of the whole was founded in these centuries?

40. In 1685, the regular infantry, and the regular cavalry of England, were respectively $\frac{3}{100}$ and $\frac{12}{1000}$ of the militia. What part were they together of the militia?

41. Of the prismatic spectrum, red occupies $\frac{1}{8}$, orange $\frac{3}{40}$, and yellow $\frac{2}{17}$. What part of the whole do these three colours occupy?

42. What part of a piece of cloth has a draper sold, who has cut off $\frac{3}{16}$, $\frac{5}{32}$, $\frac{9}{24}$, and $\frac{7}{20}$ of it?

43. A treasurer has expended $\frac{1}{10}$, $\frac{7}{18}$, $\frac{13}{48}$, $\frac{5}{80}$, and $\frac{5}{24}$ of a given sum. What part of the whole has he laid out?

44. In 1853, of the number of freshmen belonging to Cambridge $\frac{123}{412}$ belonged to Trinity College, $\frac{53}{212}$ to St John's College, $\frac{8}{38}$ to Gonville and Caius College, and $\frac{7}{212}$ to Queens' College. What part of the whole did these form?

45. Of the water of the Dead Sea, $\frac{129}{1380}$ is muriate of lime, $\frac{41}{761}$ muriate of magnesia, $\frac{59}{2380}$ muriate of soda, $\frac{1}{8080}$ sulphate of lime. What part of the whole are these ingredients?

**15.**   Let us SUBTRACT a fraction, as $\frac{3}{8}$, from an integer, as 2. We diminish one of the units by $\frac{3}{8}$; thus, $\frac{8}{8}-\frac{3}{8}=\frac{5}{8}$. This, with the other unit, makes the whole remainder $1\frac{5}{8}$. In subtracting a proper fraction from an integer, we find the *complement* of the fraction, and diminish the integer by 1.

(1) $18-\frac{13}{19}=17\frac{6}{19}.$

1. $18-\frac{4}{5}$
2. $10-\frac{6}{7}$
3. $9-\frac{4}{11}$
4. $11-\frac{9}{13}$
5. $8-\frac{2}{25}$
6. $23-\frac{23}{29}$

**15.** Let us subtract $\frac{2}{3}$ from $\frac{3}{4}$. By reducing the fractions to a common denominator, we find $\frac{2}{3} = \frac{8}{12}$, $\frac{3}{4} = \frac{9}{12}$; and by taking $\frac{8}{12}$ from $\frac{9}{12}$ we obtain the remainder $\frac{1}{12}$.

$$(2)\quad \frac{5}{7} - \frac{2}{3} = \frac{15-14}{21} = \frac{1}{21}.$$

| | | |
|---|---|---|
| 7. $\frac{1}{2} - \frac{1}{3}$ | 15. $\frac{29}{66} - \frac{3}{14}$ | 23. $\frac{3}{4} - \frac{14}{64}$ |
| 8. $\frac{1}{3} - \frac{1}{4}$ | 16. $\frac{11}{17} - \frac{7}{15}$ | 24. $\frac{11}{15} - \frac{33}{128}$ |
| 9. $\frac{2}{3} - \frac{1}{2}$ | 17. $\frac{5}{8} - \frac{4}{7}$ | 25. $\frac{3}{11} - \frac{1}{6}$ |
| 10. $\frac{4}{5} - \frac{3}{4}$ | 18. $\frac{16}{19} - \frac{13}{16}$ | 26. $\frac{14}{19} - \frac{13}{15}$ |
| 11. $\frac{3}{4} - \frac{5}{6}$ | 19. $\frac{9}{13} - \frac{5}{6}$ | 27. $\frac{16}{53} - \frac{16}{52}$ |
| 12. $\frac{11}{12} - \frac{7}{10}$ | 20. $\frac{13}{17} - \frac{9}{13}$ | 28. $\frac{19}{21} - \frac{11}{19}$ |
| 13. $\frac{5}{11} - \frac{3}{7}$ | 21. $\frac{13}{14} - \frac{12}{21}$ | 29. $\frac{14}{15} - \frac{41}{60}$ |
| 14. $\frac{11}{12} - \frac{10}{11}$ | 22. $\frac{11}{21} - \frac{17}{42}$ | 30. $\frac{19}{21} - \frac{21}{29}$ |

In subtracting a mixed number, as $2\frac{1}{2}$, from another, as $3\frac{4}{5}$, we find the difference first between the fractions, and then between the whole numbers. Thus, $\frac{1}{2}$ or $\frac{5}{10}$ from $\frac{4}{5}$ or $\frac{8}{10}$ leaves $\frac{3}{10}$, and $3 - 2 = 1$. So, $3\frac{4}{5} - 2\frac{1}{2} = 3 - 2 + \frac{8-5}{10} = 1\frac{3}{10}$.

$$(3)\quad 3\frac{5}{6} - 1\frac{9}{19} = 2 + \frac{95-54}{114} = 2\frac{41}{114}.$$

| | | |
|---|---|---|
| 31. $3\frac{1}{2} - 2\frac{1}{4}$ | 35. $18\frac{2}{3} - 10\frac{2}{5}$ | 39. $17\frac{11}{17} - 13\frac{13}{10}$ |
| 32. $7\frac{1}{3} - 5\frac{1}{5}$ | 36. $17\frac{3}{5} - 10\frac{7}{13}$ | 40. $18\frac{13}{24} - 17\frac{89}{504}$ |
| 33. $17\frac{4}{5} - 13\frac{3}{10}$ | 37. $23\frac{11}{12} - 19\frac{7}{16}$ | 41. $29\frac{11}{114} - 9\frac{13}{168}$ |
| 34. $6\frac{4}{7} - 3\frac{3}{14}$ | 38. $16\frac{13}{30} - 14\frac{23}{90}$ | 42. $3\frac{11}{31} - 2\frac{14}{186}$ |

Let us take $1\frac{2}{3}$ from $3\frac{1}{4}$. Since we cannot subtract $\frac{2}{3}$ or $\frac{4}{6}$ from $\frac{1}{2}$ or $\frac{3}{6}$, we reduce one of the 3 units to *sixths*. $1\frac{1}{2}$ or $\frac{9}{6}$ diminished by $\frac{4}{6}$ is thus $= \frac{3}{6}$. $2 - 1 = 1$. So, $3\frac{1}{2} - 1\frac{2}{3} = 2 - 1 + \frac{6+3-4}{6} = 1\frac{5}{6}$.

In subtracting the *sixths*, instead of taking 4 from 9, we may subtract 4 from 6, and add in 3. The practical advantage of this method is illustrated in A, in which we take

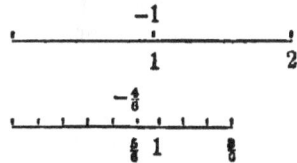

| A | | B | |
|---|---|---|---|
| $3\frac{1}{2} = 3\frac{3}{6}$ | | Units. | Sixths |
| $1\frac{2}{3} = 1\frac{4}{6}$ | | 3 " | 3 |
| $1\frac{5}{6}$ | | 1 " | 4 |
| | | 1 " | 5 |

**15.** the lower numerator from the common denominator, and add in the upper numerator. In B, we may consider the *sixths* as units of a lower name, of which six make up a higher unit. The solution is then obtained as in Compound Subtraction.

$$(4)\ 6\tfrac{1}{2} - 3\tfrac{17}{18} = 5 - 3 + \tfrac{18 + 9 - 17}{18} = 2\tfrac{10}{18} = 2\tfrac{5}{9}.$$

| | | |
|---|---|---|
| 43. $20\tfrac{1}{2} - 16\tfrac{7}{8}$ | 47. $14\tfrac{4}{9} - 13\tfrac{9}{10}$ | 51. $13\tfrac{4}{8} - 10\tfrac{17}{21}$ |
| 44. $16\tfrac{1}{2} - 13\tfrac{4}{5}$ | 48. $15\tfrac{11}{16} - 4\tfrac{13}{18}$ | 52. $6\tfrac{9}{10} - 5\tfrac{19}{21}$ |
| 45. $14\tfrac{2}{3} - 9\tfrac{11}{17}$ | 49. $13\tfrac{7}{9} - 11\tfrac{4}{5}$ | 53. $18\tfrac{13}{17} - 11\tfrac{10}{13}$ |
| 46. $16\tfrac{7}{9} - 10\tfrac{3}{5}$ | 50. $8\tfrac{1}{2} - 4\tfrac{11}{21}$ | 54. $23\tfrac{23}{25} - 22\tfrac{13}{13}$ |

(5) $\tfrac{2}{5}$ of a pole is below the level of a pond, $\tfrac{7}{20}$ of it is in the water. How much of it is in the ground?

$$\tfrac{2}{5} - \tfrac{7}{20} = \tfrac{12 - 7}{20} = \tfrac{5}{20} = \tfrac{1}{4}.$$

55. $\tfrac{2}{3}$ of a pole is above the bottom of a pool, and $\tfrac{3}{14}$ is in the pool. What part of it is above the level of the water?

56. A retail draper who has bought $\tfrac{3}{4}$ of a piece of cloth, sells $\tfrac{1}{16}$ of the piece. What part of it has he over?

57. $\tfrac{2}{3}$ of a common is laid out as bleaching-ground. What part of it is over?

58. A sailor has spent $\tfrac{8}{13}$ of his life at sea. What part was spent before he went to sea?

59. A person succeeding to a legacy left by an ancestor or descendant in the direct line, pays $\tfrac{1}{100}$ of the value as duty. What part is over?

60. Of the prismatic spectrum, the blue, indigo, and violet rays together occupy $\tfrac{1}{2}$, and the blue and indigo together $\tfrac{5}{18}$. What part does the violet occupy?

61. The number of pear and apple trees in an orchard is $\tfrac{1}{2}$ of that of the whole, and that of the pear trees is $\tfrac{7}{32}$. What part of the whole is the number of apple trees?

62. Of a consignment of guano from Saldanha Bay, $\tfrac{201}{408}$ consisted of carbonate of lime and phosphates of lime and magnesia, and $\tfrac{18}{48}$ of the phosphates. What part of it was carbonate of lime?

63. In 1857, the number of parliamentary electors in Scotland was $\tfrac{87}{977}$ of the whole number in Great Britain. What part of the whole number was the number in England and Wales?

**16.**   (1) $\tfrac{1}{2} + \tfrac{2}{3} - \tfrac{3}{4} + \tfrac{4}{5} - \tfrac{3}{10} = \tfrac{30 + 40 - 45 + 48 - 18}{60} = \tfrac{55}{60} = 1\tfrac{1}{4}.$

| | |
|---|---|
| 1. $\tfrac{1}{2} + \tfrac{2}{3} - \tfrac{3}{4} - \tfrac{3}{8}$ | 4. $\tfrac{5}{9} + \tfrac{3}{5} - \tfrac{4}{15} + \tfrac{1}{2}$ |
| 2. $\tfrac{2}{3} - \tfrac{1}{8} + \tfrac{5}{6} - \tfrac{1}{2}$ | 5. $\tfrac{7}{8} + \tfrac{3}{4} + \tfrac{9}{10} - 1\tfrac{4}{5}$ |
| 3. $\tfrac{7}{8} + \tfrac{5}{6} - \tfrac{3}{4} - \tfrac{1}{48}$ | 6. $\tfrac{3}{7} + \tfrac{9}{14} - \tfrac{5}{21} + \tfrac{7}{8}$ |

**16.** 7. $2\frac{2}{3} + 3\frac{1}{2} - \frac{7}{8} - \frac{11}{16} + \frac{3}{5}$    10. $\frac{11}{12} - \frac{7}{8} + \frac{4}{15} - \frac{13}{16} + 1\frac{11}{80}$

8. $3\frac{1}{4} - \frac{3}{7} + 2\frac{2}{5} - 4\frac{1}{2} + \frac{3}{8}$    11. $5\frac{5}{8} - \frac{11}{20} + \frac{4}{9} - \frac{3}{8} - 2\frac{1}{13}$

9. $2\frac{1}{4} - \frac{3}{8} + 2\frac{2}{3} - \frac{7}{8} + \frac{11}{14}$    12. $4\frac{1}{3} - \frac{7}{8} + 3\frac{1}{4} - \frac{11}{10} + \frac{7}{15}$

(2) A traveller has gone $\frac{1}{9}$ of a journey on foot, $\frac{2}{15}$ on horseback, $\frac{1}{4}$ by rail, and the rest by coach. What part has he gone by coach?

$$1 - \left(\frac{1}{9} + \frac{2}{15} + \frac{1}{4}\right) = 1 - \left(\frac{20 + 24 + 45}{180}\right)$$
$$= 1 - \frac{89}{180} = \frac{91}{180}.$$

13. $\frac{1}{4}$ of a pole is blue, $\frac{2}{7}$ red, and the rest white. What part of it is white?

14. A student has in three weeks read respectively $\frac{4}{21}$, $\frac{2}{7}$, and $\frac{1}{4}$ of the First Book of the Æneid. What part of it has he yet to read?

15. A soldier while in the army had spent $\frac{1}{4}$ of his life in the United Kingdom, $\frac{6}{17}$ in Canada, $\frac{1}{18}$ in Gibraltar, $\frac{1}{8}$ in India, and $\frac{1}{17}$ in the Crimea. What part of his life had he spent before enlisting?

16. A jeweller having used $\frac{3}{10}$, $\frac{7}{50}$, and $\frac{11}{45}$ of an ingot of gold, wishes to know what part still remains.

17. Of the whole time spent by Professor Piazzi Smyth in the Astronomical Expedition to Teneriffe in 1856, $\frac{2}{13}$ was spent in the lowlands of Teneriffe, $\frac{4}{117}$ at Guajara, and $\frac{94}{117}$ at Alta Vista. What part was spent in the voyage?

18. Of the component elements of albumen, $\frac{11}{15}$ is carbon, $\frac{7}{100}$ hydrogen, and $\frac{6}{38}$ nitrogen. What part of it does the remainder, consisting of oxygen, phosphorus, &c., constitute?

19. Of the whole number of Jehoshaphat's " men of valour " in Judah and Benjamin, the three divisions of Judah were respectively $\frac{6}{28}$, $\frac{7}{28}$, and $\frac{6}{28}$. What part belonged to Benjamin?

20. Of the black and mulatto population of Cuba in 1850, the free mulattoes were $\frac{444}{2125}$, the free blacks $\frac{874}{2125}$, and the mulatto slaves $\frac{218}{2125}$. What part was the number of black slaves?

21. Of the annual salaries of the principal, depute, and assistant clerks of the Court of Session, 5 deputes receive $\frac{8}{183}$ each, and 9 assistants $\frac{7}{183}$ each. What part does each of the 4 principals receive?

**17.** In MULTIPLYING a fraction by another, as $\frac{4}{5}$ by $\frac{2}{3}$, we consider that since the multiplier $\frac{2}{3}$ is $\frac{1}{3}$ of 2, the product will be $\frac{1}{3}$ of 2 times $\frac{4}{5}$. Now $2 \times \frac{4}{5} = \frac{8}{5}$, and the required product is $= \frac{8}{5} \div 3 = \frac{8}{15}$, which is thus $= \frac{4}{5} \times \frac{2}{3}$.

In multiplying fractions together, the product of the numer-

**17.** ators becomes the numerator of the product, and the product of the denominators the denominator of the product.

In multiplying by an integer we repeat the multiplicand as many times as there are units in the multiplier; in multiplying by a fraction we take that part of the multiplicand which is denoted by the multiplier.

$\frac{4}{5} \times \frac{3}{5}$ may be expressed as $\frac{3}{5}$ of $\frac{4}{5}$, or $\frac{4}{5}$ of $\frac{3}{5}$, which being the fraction of a fraction is termed a COMPOUND FRACTION, in contradistinction to a SIMPLE FRACTION, as $\frac{4}{5}$. A Compound Fraction is reduced to the form of a simple one by multiplying the numerators and the denominators, as in Multiplication of Fractions; thus, $\frac{2}{3}$ of $\frac{4}{5} = \frac{8}{15}$.

We may consider $\frac{2}{3}$ of $\frac{4}{5}$ either as $\frac{1}{3}$ of $2 \times \frac{4}{5}$, or, as in the diagram, we may divide $\frac{4}{5}$ into *three* equal parts, and take *two* of them. Similarly, we may take $\frac{4}{5}$ of $\frac{2}{3}$, either as $\frac{1}{5}$ of $4 \times \frac{2}{3}$, or, as in the diagram, we may divide $\frac{2}{3}$ into *five* equal parts and take *four* of them.

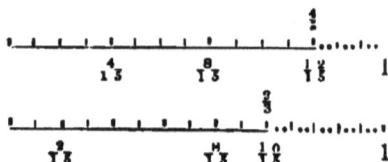

$$(1)\quad \frac{8}{9} \times 1\frac{7}{10} = \frac{8}{9} \times \frac{17}{10} = \frac{68}{45} = 1\frac{23}{45}.$$

Since 2 is a common factor of 8 and 10, we CANCEL these numbers, and write the number of times the factor is contained in each. By thus cancelling any numerator with any denominator with which it has a common factor, we obtain the product in its lowest terms.

$$(2)\quad \frac{6}{35} \times \frac{7}{18} = \frac{6}{35} \times \frac{7}{18} = \frac{1}{15}.$$

The numerator of the product is $= 1 \times 1$. *Unity* takes the place of a numerator or a denominator cancelled with any of its multiples.

$$(3)\quad 6\frac{2}{3} \times \frac{31}{50} = \frac{20}{3} \times \frac{31}{50} = \frac{62}{15} = 4\frac{2}{15}.$$

| | | |
|---|---|---|
| 1. $\frac{3}{4} \times \frac{4}{5}$ | 9. $\frac{3}{4} \times 5\frac{3}{4}$ | 17. $9\frac{1}{7} \times 8\frac{3}{64}$ |
| 2. $\frac{3}{8} \times \frac{7}{11}$ | 10. $\frac{4}{7} \times 7\frac{2}{3}$ | 18. $7\frac{3}{13} \times \frac{200}{217}$ |
| 3. $\frac{5}{6} \times \frac{4}{7}$ | 11. $\frac{3}{5} \times 16\frac{1}{4}$ | 19. $19\frac{1}{2} \times 16\frac{112}{193}$ |
| 4. $\frac{3}{7} \times \frac{14}{27}$ | 12. $\frac{2}{3} \times 18\frac{1}{5}$ | 20. $23\frac{1}{4} \times 3\frac{1}{47}$ |
| 5. $\frac{11}{13} \times \frac{26}{77}$ | 13. $1\frac{3}{11} \times \frac{11}{13}$ | 21. $17\frac{3}{4} \times \frac{139}{142}$ |
| 6. $\frac{3}{4} \times \frac{32}{33}$ | 14. $2\frac{9}{11} \times \frac{4}{7}$ | 22. $\frac{1480}{1386} \times \frac{111}{343}$ |
| 7. $\frac{7}{15} \times \frac{44}{64}$ | 15. $\frac{4}{5} \times 6\frac{4}{5}$ | 23. $4\frac{1}{11} \times 17\frac{1}{10}$ |
| 8. $\frac{16}{19} \times \frac{79}{81}$ | 16. $\frac{3}{11} \times 7\frac{1}{4}$ | 24. $\frac{113}{363} \times 3\frac{177}{1250}$ |

**17.**  (4)  $\frac{3}{7} \times 5\frac{2}{3} \times 4\frac{5}{11} = \frac{3}{7} \times \frac{17}{3} \times \frac{\overset{7}{\cancel{49}}}{11} = \frac{119}{11} = 10\frac{9}{11}.$

| | |
|---|---|
| 25. $\frac{1}{2} \times 2\frac{2}{3} \times \frac{3}{4}$ | 31. $\frac{3}{8} \times \frac{19}{33} \times 8\frac{1}{4}$ |
| 26. $\frac{1}{4} \times 3\frac{1}{4} \times \frac{7}{8}$ | 32. $6\frac{3}{4} \times \frac{8}{27} \times 6\frac{2}{3}$ |
| 27. $3\frac{3}{4} \times \frac{12}{15} \times \frac{13}{32}$ | 33. $\frac{4}{5} \times \frac{4}{7} \times 6\frac{3}{4}$ |
| 28. $2\frac{1}{4} \times \frac{1}{8} \times 3\frac{5}{8}$ | 34. $\frac{9}{11} \times \frac{3}{7} \times 4\frac{1}{4}$ |
| 29. $3\frac{1}{11} \times \frac{1}{17} \times 5\frac{1}{4}$ | 35. $7\frac{1}{4} \times \frac{2}{5} \times \frac{11}{29}$ |
| 30. $3\frac{2}{3} \times \frac{9}{58} \times 6\frac{1}{4}$ | 36. $8\frac{1}{8} \times 1\frac{3}{7} \times \frac{21}{215}$ |

Reduce the following *Compound* to Simple Fractions:—

(5)  $\frac{1}{4}$ of $\frac{3}{7}$ of $\frac{5}{11}$ of $\frac{22}{25} = \frac{1}{\underset{2}{\cancel{4}}} \times \frac{3}{7} \times \frac{5}{\cancel{11}} \times \frac{\overset{2}{\cancel{22}}}{\underset{5}{\cancel{25}}} = \frac{3}{70}.$

(6)  $\frac{2}{5}$ of $\frac{3}{7}$ of $7\frac{1}{2}$ of $\frac{3}{5} = \frac{2}{\cancel{5}} \times \frac{3}{7} \times \frac{\overset{3}{\cancel{15}}}{\cancel{2}} \times \frac{\overset{4}{\cancel{36}}}{\underset{11}{\cancel{55}}} = \frac{36}{77}.$

| | |
|---|---|
| 37. $\frac{1}{2}$ of $\frac{3}{4}$ of $\frac{4}{5}$ | 46. $\frac{13}{17}$ of $\frac{19}{280}$ of $4\frac{1}{4}$ of $\frac{20}{381}$ |
| 38. $\frac{1}{5}$ of $\frac{3}{8}$ of $1\frac{3}{5}$ | 47. $\frac{1}{4}$ of $\frac{3}{5}$ of $\frac{5}{6}$ of $\frac{7}{8}$ of $\frac{7}{9}$ of 7 |
| 39. $\frac{3}{5}$ of $2\frac{3}{4}$ of $\frac{6}{13}$ | 48. $\frac{3}{4}$ of $5\frac{1}{2}$ of $\frac{3}{33}$ of $\frac{19}{27}$ of 5 |
| 40. $\frac{3}{4}$ of $\frac{5}{8}$ of $1\frac{1}{8}$ | 49. $\frac{22}{29}$ of $16\frac{2}{3}$ of $\frac{3}{29}$ of $70\frac{1}{13}$ |
| 41. $\frac{7}{8}$ of $\frac{3}{4}$ of $8\frac{1}{2}$ | 50. $\frac{13}{14}$ of $\frac{80}{189}$ of $\frac{17}{48}$ of $\frac{11}{120}$ of $307\frac{3}{11}$ |
| 42. $\frac{3}{8}$ of $1\frac{1}{4}$ of $\frac{7}{9}$ of $4\frac{1}{2}$ | 51. $\frac{1}{8}$ of $\frac{1}{17}$ of $\frac{3}{33}$ of $8\frac{1}{4}$ of $8\frac{1}{2}$ |
| 43. $\frac{5}{8}$ of $2\frac{1}{4}$ of $\frac{3}{7}$ of $5\frac{1}{4}$ | 52. $\frac{1}{3}$ of $\frac{1}{9}$ of $\frac{1}{81}$ of $\frac{1}{729}$ of 6561 |
| 44. $\frac{3}{8}$ of $\frac{7}{13}$ of $1\frac{2}{8}$ of $3\frac{1}{4}$ of $1\frac{3}{8}$ | 53. $\frac{3}{11}$ of 19 of $\frac{4}{35}$ of $4\frac{1}{4}$ of $1\frac{1}{7}$ of $8\frac{2}{3}$ |
| 45. $\frac{9}{50}$ of $\frac{5}{8}$ of $4\frac{11}{12}$ of $3\frac{2}{3}$ | 54. $\frac{3}{17}$ of $3\frac{4}{8}$ of $7\frac{1}{5}$ of $7\frac{1}{13}$ of $\frac{60}{91}$ |

(7)  If a train runs $\frac{4}{5}$ of a mile in a minute; how many miles will it run in $\frac{3}{8}$ of $43\frac{1}{2}$ min. ?

ml.

$\frac{4}{5} \times \frac{3}{8} \times 43\frac{1}{2} = \frac{\overset{}{\cancel{4}}}{\underset{\cancel{8}}{\cancel{5}}} \times \frac{3}{\underset{2}{\cancel{5}}} \times \frac{\overset{29}{\cancel{87}}}{2} = \frac{29}{4} = 7\frac{1}{4}$ ml.

55. A soldier was in hospital $\frac{3}{18}$ of the time he served in India, which was $\frac{6}{13}$ of his life. What part of his life was he in hospital?

56. A sailor's share of prize-money is $\frac{7}{10}$ of a midshipman's, whose share is $\frac{3}{4}$ of a lieutenant's. What part of a lieutenant's share does a sailor get?

57. Jack, who gets $\frac{1}{3}$ of a plum-pudding, gives $\frac{3}{5}$ of his share to Tom, who gives $\frac{1}{2}$ of his to Harry. What part of the plum-pudding does Harry get?

**17.**  58. A schoolboy prepares his lessons at home in $\frac{4}{5}$ of the time he plays, which amounts to $\frac{3}{10}$ of $\frac{5}{8}$ of a day. During what part of a day does he prepare his lessons?

59. On the Geelong and Melbourne Railway, the fare per mile by the third class is $\frac{7}{12}$ of that by the second, which is $\frac{4}{5}$ of that by the first, which is $3\frac{3}{4}$d. Find the fare per mile by the first.

60. Find the receipts of a railway for a week which amount to $7\frac{5}{8}\frac{1}{5}$ of £6384.

61. The number of registrars employed in the Census of 1851 was $\frac{109}{1887}$ of that of the enumerators, of whom there were 38740. Find the number of registrars.

62. If a train runs a mile in $\frac{4}{5}$ of $3\frac{3}{4}$ min., in what time will it run $\frac{7}{10}$ of $23\frac{1}{2}$ miles?

63. 24 flagstaffs are placed on a road at the distance of $\frac{4}{5}$ of $73\frac{1}{3}$ yards between each. How many yards are between the first and the last.

The number of spaces between a number of objects placed in a line is *one* less than the number of objects.

**18.**  In DIVIDING a fraction by another, as $\frac{3}{4}$ by $\frac{2}{3}$, we consider that since the divisor $\frac{2}{3}$ is $\frac{1}{3}$ of 2, the quotient obtained by dividing by $\frac{2}{3}$ is 3 times as large as that obtained by dividing by 2. Now $\frac{3}{4} \div 2 = \frac{3}{8}$, and the required quotient is $= \frac{3}{8} \times 3 = \frac{9}{8}$. $\frac{3}{4} \div \frac{2}{3}$ thus produces the same result as $\frac{3}{4} \times \frac{3}{2}$.

In dividing a fraction by another, we invert the divisor, and proceed as in Multiplication of Fractions. A fraction inverted is the RECIPROCAL of the original fraction; thus, $\frac{3}{2}$ is the reciprocal of $\frac{2}{3}$. The product of a fraction by its reciprocal is = unity.

To divide $\frac{3}{4}$ by $\frac{2}{3}$, we may, as in the first diagram, according to the previous explanation, take *one-half* of $\frac{3}{4}$, which is $\frac{3}{8}$, and by taking *three* parts each $= \frac{3}{8}$, we obtain $\frac{9}{8}$.

Expressing $\frac{3}{4}$ and $\frac{2}{3}$ in the same denominator as $\frac{9}{12}$ and $\frac{8}{12}$ respectively, we see in the second diagram that if we take 8 *twelfths* as the unit, 9 *twelfths* contain 9 of those parts of which the unit contains 8. $\frac{9}{12}$ is thus $\frac{9}{8}$ of $\frac{8}{12}$, or $\frac{9}{8}$ is the quotient obtained by dividing $\frac{9}{12}$ by $\frac{8}{12}$, or is $= \frac{3}{4} \div \frac{2}{3}$.

(1)  $\frac{4}{5} \div \frac{7}{11} = \frac{4}{5} \times \frac{11}{7} = \frac{44}{35} = 1\frac{9}{35}.$

(2)  $4\frac{3}{5} \div 7\frac{11}{15} = \frac{23}{5} \times \frac{\overset{3}{45}}{116} = \frac{69}{116}.$

**18.**

| | | |
|---|---|---|
| 1. $\frac{2}{4} \div \frac{4}{8}$ | 9. $\frac{4}{13} \div \frac{5}{26}$ | 17. $19\frac{3}{8} \div \frac{31}{52}$ |
| 2. $\frac{4}{6} \div \frac{3}{5}$ | 10. $\frac{17}{61} \div \frac{34}{100}$ | 18. $17\frac{4}{17} \div \frac{38}{39}$ |
| 3. $\frac{1}{4} \div \frac{2}{4}$ | 11. $\frac{10}{41} \div \frac{13}{42}$ | 19. $4\frac{2}{3} \div 5\frac{1}{4}$ |
| 4. $\frac{7}{9} \div \frac{11}{18}$ | 12. $\frac{33}{47} \div \frac{31}{41}$ | 20. $5\frac{5}{8} \div 11\frac{3}{4}$ |
| 5. $\frac{14}{17} \div \frac{15}{81}$ | 13. $5\frac{3}{4} \div \frac{11}{18}$ | 21. $2\frac{17}{19} \div 7\frac{6}{9}$ |
| 6. $\frac{17}{16} \div \frac{34}{39}$ | 14. $7\frac{1}{2} \div \frac{16}{19}$ | 22. $3\frac{11}{12} \div 9\frac{2}{5}$ |
| 7. $\frac{11}{4} \div \frac{23}{34}$ | 15. $3\frac{3}{4} \div \frac{7}{8}$ | 23. $14\frac{2}{3} \div 4\frac{5}{9}$ |
| 8. $\frac{14}{17} \div \frac{10}{81}$ | 16. $6\frac{1}{2} \div \frac{11}{14}$ | 24. $7\frac{3}{4} \div 2\frac{1}{8}$ |

(3) $\frac{7}{8} \div \frac{7}{11}$ of $3\frac{1}{3} = \frac{\overset{7}{\cancel{7}}}{\underset{8}{\cancel{8}}} \times \frac{11}{\cancel{7}} \times \frac{\cancel{9}}{10} = \frac{11}{10}.$

(4) $\frac{3}{37}$ of $4\frac{1}{4} \div \frac{4}{5} = \frac{3}{37} \times \frac{17}{\underset{}{\cancel{4}}} \times \frac{\overset{2}{\cancel{8}}}{5} = \frac{102}{185}.$

| | |
|---|---|
| 25. $\frac{21}{4} \div \frac{3}{4}$ of $10\frac{1}{3}$ | 31. $\frac{1}{4}$ of $\frac{14}{15} \div \frac{7}{28}$ |
| 26. $\frac{181}{154} \div \frac{7}{9}$ of $25\frac{7}{8}$ | 32. $\frac{1}{4}$ of $2\frac{3}{4} \div \frac{1}{7}$ |
| 27. $\frac{41}{45} \div \frac{3}{8}$ of $12\frac{2}{3}$ | 33. $\frac{3}{7}$ of $1\frac{7}{16} \div \frac{31}{43}$ of $\frac{53}{81}$ |
| 28. $\frac{19}{24} \div \frac{4}{9}$ of $3\frac{1}{4}$ | 34. $\frac{4}{7}$ of $\frac{23}{30} \div \frac{9}{73}$ of 4 |
| 29. $3\frac{2}{3} \div \frac{18}{24}$ of $7\frac{1}{4}$ | 35. $\frac{9}{10}$ of $\frac{11}{12} \div \frac{3}{4}$ of $1\frac{1}{11}$ |
| 30. $\frac{11}{14} \div \frac{4}{5}$ of $8\frac{7}{16}$ | 36. $\frac{3}{7}$ of $\frac{48}{49} \div \frac{3}{8}$ of $\frac{7}{13}$ |

We may write the quotient $\frac{3}{4} \div \frac{4}{5}$ in the following form :—
$\frac{\frac{3}{4}}{\frac{4}{5}}$, in which the dividend becomes the numerator, and the divisor the denominator of a COMPLEX FRACTION.

A Complex Fraction has a fraction in either its numerator or denominator, or in both of them :—thus, $\frac{\frac{3}{4}}{\frac{4}{5}}$, $\frac{2}{7\frac{1}{4}}$, $\frac{5\frac{1}{2}}{9}$, $\frac{3\frac{3}{4}}{7\frac{3}{8}}$, are complex fractions. The reduction of Complex to Simple Fractions is similar to the Division of Fractions.

Reduce the following *Complex* to Simple Fractions :—

(5) $\frac{\frac{3}{4}}{7} = \frac{3}{28}.$

We have multiplied the numerator and the denominator of the fraction by 4, the denominator of the numerator. So, when either the numerator or the denominator is an integer, we multiply the numerator and the denominator by *the denominator of the fractional term*.

(6) $\frac{\frac{3}{5}}{1\frac{1}{3}} = \frac{3}{5} \div 1\frac{1}{3} = \frac{3}{5} \times \frac{3}{4} = \frac{9}{20}.$

or, $\left[ \frac{\frac{3}{5}}{1\frac{1}{3}} \right] = \frac{9}{20}.$

**18.**  The quotient may evidently be obtained by multiplying the extremes of the complex fraction for the numerator, and the means for the denominator.

$$(7)\quad \frac{8\frac{3}{4}}{12\frac{5}{6}} = \left[\begin{array}{c} 5 \ \frac{\cancel{35}}{\cancel{4}}\ ^2 \\ \frac{\cancel{77}}{\cancel{6}}\ _{11} \\ 3 \end{array}\right] = \frac{14}{22}.$$

We may cancel either of the extremes with either of the means. As the numerator and the denominator will likely be expressed in the lowest terms, we thus cancel the *first* with the *third*, as 35 with 77, and the *second* with the *fourth*, as 4 with 6.

| | | |
|---|---|---|
| 37. $\dfrac{6}{8\frac{1}{4}}$ | 42. $\dfrac{\frac{3}{8}}{\frac{4}{5}}$ | 47. $\dfrac{1\frac{3}{14}}{7\frac{4}{5}}$ |
| 38. $\dfrac{7}{11\frac{3}{4}}$ | 43. $\dfrac{\frac{4}{9}}{\frac{7}{10}}$ | 48. $\dfrac{5\frac{3}{4}}{6\frac{2}{3}}$ |
| 39. $\dfrac{9}{12\frac{5}{8}}$ | 44. $\dfrac{\frac{10}{11}}{\frac{12}{13}}$ | 49. $\dfrac{19\frac{3}{4}}{28\frac{7}{16}}$ |
| 40. $\dfrac{11}{13\frac{4}{7}}$ | 45. $\dfrac{\frac{9}{7}}{\frac{14}{13}}$ | 50. $\dfrac{2\frac{8}{11}}{7\frac{3}{4}}$ |
| 41. $\dfrac{4\frac{7}{9}}{7}$ | 46. $\dfrac{\frac{16}{19}}{\frac{45}{76}}$ | 51. $\dfrac{23\frac{7}{9}}{24\frac{3}{8}}$ |

$$(8)\quad \frac{\frac{3}{8}\text{ of }3\frac{1}{4}}{\frac{4}{5}\text{ of }3\frac{4}{7}} = \frac{\frac{3}{\cancel{8}}\times\frac{13}{\cancel{4}}\,^2}{\frac{4}{\cancel{5}}\times\frac{\cancel{25}}{7}\,^5} = \frac{13\times7}{6\times20} = \frac{91}{120}.$$

$$(9)\quad \frac{5\frac{9}{14}}{\frac{3}{11}} = \frac{\frac{79}{14}}{\frac{3}{11}} = \frac{\frac{869}{42}}{5} = \frac{869}{210} = 4\frac{29}{210}.$$

$$(10)\quad \frac{3}{\dfrac{\frac{4}{7}}{\frac{9}{11}}} = \frac{3}{\frac{44}{21}} = \frac{63}{44} = 1\frac{19}{44}.$$

| | | |
|---|---|---|
| 52. $\dfrac{\frac{3}{4}\text{ of }3\frac{1}{2}}{\frac{4}{5}\text{ of }9\frac{2}{3}}$ | 54. $\dfrac{\frac{2}{11}\text{ of }12\frac{4}{7}}{\frac{3}{8}\text{ of }\frac{5}{6\frac{2}{3}}}$ | 56. $\dfrac{\frac{2}{3}}{\dfrac{\frac{3}{4}}{5}}$ |
| 53. $\dfrac{\frac{3}{8}\text{ of }13\frac{1}{3}}{\frac{4}{7}\text{ of }7\frac{7}{8}}$ | 55. $\dfrac{\frac{3}{4}\text{ of }\frac{2}{7}}{\frac{4}{7\frac{1}{2}}\text{ of }\frac{8}{11}}$ | 57. $\dfrac{2}{\dfrac{\frac{3}{4}}{\frac{5}{6}}}$ |

VULGAR FRACTIONS. 37

**18.**  (11) How many pieces, each $30\frac{3}{4}$ yards, are contained in $114\frac{4}{5}$ yards ?

$$114\frac{4}{5} \div 30\frac{3}{4} = \frac{\overset{14}{\cancel{574}}}{5} \times \frac{4}{\underset{3}{\cancel{123}}} = \frac{56}{15} = 3\frac{11}{15} \text{ pieces.}$$

58. If a piece of cloth is $29\frac{3}{4}$ yards in length, and a remnant $1\frac{5}{8}$ yard; how many times is the former as long as the latter?

59. How many squares, each $\frac{1}{8}$ sq. inch, are contained in $132\frac{1}{4}$ sq. inches?

60. How many postage-stamps, containing $\frac{4}{8}$ sq. in., are in a sheet of $172\frac{1}{3}$ sq. in. ?

61. How many times can a measure of $\frac{7}{8}$ pint be filled out of a vessel containing $63\frac{1}{4}$ pints?

62. How many times will a coin $2\frac{1}{4}$ inches in circumference turn round in traversing 30 inches?

63. Mercury is $13\frac{3}{5}$ times as heavy as water, and gold is $19\frac{3}{8}$ times. How many times is gold as heavy as mercury?

64. The pellicle from which goldbeaters' skin is made is $\frac{1}{3000}$ inch thick, while gold leaf is $\frac{1}{282500}$ in. thick. How many times is the former as thick as the latter?

65. The largest scale of the Ordnance Survey Maps is lineally $\frac{1}{2172}$ of that of nature, and the smallest is $\frac{1}{3380}$. How many times is the former as large as the latter?

66. The mass of the Earth is $\frac{1}{335551}$, and that of Jupiter is $\frac{1}{1048}$ of that of the Sun. How many times is the mass of Jupiter as great as that of the Earth?

67. A book of 240 leaves without boards is $\frac{11}{16}$ inch thick, and another of 180 leaves without boards is $\frac{7}{10}$ inch thick. How many times is the paper of the former as thick as that of the latter?

(12) How many men are in a regiment of which $\frac{3}{10} = $ 255 men ?

$$255 \div \tfrac{3}{10} = 255 \times \tfrac{10}{3} = 850 \text{ men.}$$

The regiment is evidently $= \frac{10}{3}$ of $\frac{3}{10}$ of the regiment, but $\frac{3}{10}$ of the regiment $= 255$; hence the number in the regiment $= \frac{10}{3}$ of $255 = 850$.

68. Find the length of a pole of which $\frac{3}{4} = 18$ ft.

69. The *Pylades* war steamer, having 21 guns, has $\frac{7}{13}$ the number which the *Princess Royal* war steamer has. Find the number of guns in the latter.

70. Find the distance from London to Kurrachee, that from the head of the Red Sea to Kurrachee, which is 1700 miles, being $\frac{11}{13}$ of it.

**18.** (13) Of a pole, $\frac{1}{12}$ is painted white, $\frac{7}{30}$ green, $\frac{11}{60}$ red, and the remainder which is 5 ft. is painted black. Find the length of the pole.

$$\tfrac{1}{12}+\tfrac{7}{30}+\tfrac{11}{60}=\tfrac{5+14+11}{60}=\tfrac{30}{60}=\tfrac{1}{2}$$
$$1-\tfrac{1}{2}=\tfrac{1}{2};\ 5\text{ ft.}\div\tfrac{1}{2}=10\text{ ft.}$$

71. Of the area of the five great lakes, Lakes Erie and Ontario together contain $\frac{1}{4}$, Michigan and Huron together $\frac{19}{44}$, while Lake Superior contains 32000 sq. miles. How many square miles do they in all contain?

72. Of an army $\frac{1}{3}$ is English, $\frac{3}{7}$ Scotch, $\frac{7}{150}$ Welsh, and the remainder numbers 4796 Irish. How many are there in all?

73. Of the distance from Edinburgh to London by rail, via Carlisle, that from Edinburgh to Carlisle is $\frac{1}{4}$, from Carlisle to Preston $\frac{9}{50}$, while that from Preston to London is 210 miles. Find the distance from Edinburgh to London.

(14) A labourer can do a piece of work in $18\frac{4}{5}$ days. What part of it can he do in a day?

$$1\div 18\tfrac{4}{5}=\tfrac{5}{94}.$$

74. A labourer can perform a piece of work in $12\frac{1}{2}$ days. What part of it can he do in a day?

75. A workman can floor a room in $5\frac{11}{13}$ days. What part of the room can he floor in a day?

(15) A can do a work in 8 days, B in 12 da., and C in 16 da. In what time will they do it working together?

A can do $\frac{1}{8}$ of the work in 1 day.
B 〃 $\frac{1}{12}$ 〃 〃
C 〃 $\frac{1}{16}$ 〃 〃

A, B, and C can do $\frac{1}{8}+\frac{1}{12}+\frac{1}{16}=\frac{6+4+3}{48}=\frac{13}{48}$ of the work in 1 da. A, B, and C, will thus do the *whole* work in as many days as are $=1\div\frac{13}{48}=\frac{48}{13}=3\frac{9}{13}$ da.

(16) A can do a work in $10\frac{1}{2}$ da., B in $12\frac{1}{4}$ da., and C in $8\frac{3}{4}$ da. In what time will they together do it?

$$\text{A }10\tfrac{1}{2}=\tfrac{21}{2}\quad \tfrac{2}{21}=\tfrac{70}{735}$$
$$\text{B }12\tfrac{1}{4}=\tfrac{49}{4}\quad \tfrac{4}{49}=\tfrac{60}{735}$$
$$\text{C }8\tfrac{3}{4}=\tfrac{35}{4}\quad \tfrac{4}{35}=\tfrac{84}{735}$$
$$\tfrac{214}{735};\ \tfrac{735}{214}=3\tfrac{93}{214}.$$

**18.** 76. D can do a work in 6 da., E in 9 da., and F in 10 da. In what time will they do it by working together?

77. A cistern can be filled by three pipes in 10, 12, and 18 min. respectively. In what time will it be filled when they are all open?

78. X can do a work in 3 hours, Y in $4\frac{1}{2}$ ho., and Z in $6\frac{3}{4}$ ho. In what time will they together do it?

79. A can do a work in $10\frac{1}{2}$ da., B in $11\frac{1}{2}$, and C in $12\frac{1}{2}$. In what time will they do it together?

80. A can do a work in 3 da., B in 4 da., and C can do as much as A and B together. In what time will they do it working together?

☞ C can do $\frac{1}{3} + \frac{1}{4}$ of the work in a day.

81. A can do a work in 7 hours, B in $5\frac{1}{4}$ hours, and C can work twice as fast as A. In what time will they do it together?

82. A, B, C, can do a work together in 20 days; A alone can do it in 40 da., B alone in 60 da. In what time can C alone do it?

☞ C can do $\frac{1}{20} - (\frac{1}{40} + \frac{1}{60})$ of the work in a day.

83. D, E, F, can do a work together in 5 days, D in $16\frac{2}{3}$, and E in $13\frac{1}{3}$ da. In what time can F alone do it?

84. A, B, C, can do a work together in 7 days, which A and B can do together in 10 da. In what time will C do it?

☞ C can do $\frac{1}{7} - \frac{1}{10}$ of the work in a day.

85. F, G, H, can perform a work together in 1 day, which G and H can do together in $1\frac{1}{2}$ day. In what time can F do it?

86. X and Y can accomplish a work together in 8 days, Y and Z together in 9 da., and Y in 14 da. In what time can X and Z do it separately and together?

☞ X can do $\frac{1}{8} - \frac{1}{14}$ of the work in a day.

87. A and B can do a work together in $3\frac{1}{3}$ da., B and C together in 4 da., and B in $5\frac{1}{3}$ da. In what time can A and C do it separately and together?

**19.** (1) $\frac{2}{13}$ of $3\frac{19}{24} + \dfrac{11\frac{1}{5}}{18\frac{2}{7}}$.

$$\frac{2}{13} \text{ of } 3\frac{19}{24} = \frac{2}{13} \times \frac{91}{24} = \frac{7}{12} = \frac{140}{240}$$

$$\dfrac{11\frac{1}{5}}{18\frac{2}{7}} = \frac{56}{5} \times \frac{7}{128} = \frac{49}{80} = \frac{147}{240}$$

$$\frac{287}{240} = 1\frac{47}{240}.$$

1. $\frac{3}{4}$ of $\frac{7}{9} + \frac{5}{6}$ of $\frac{2}{3} + \frac{3}{8}$ of $1\frac{3}{4}$

2. $\frac{5}{6}$ of $3\frac{1}{4} + \frac{3}{5}$ of $\frac{7}{9} + \frac{2}{11}$ of $3\frac{2}{3}$

3. $\frac{2}{5}$ of $4\frac{1}{2} + \frac{3}{4}$ of $\frac{7}{11} + \frac{10}{21}$ of $1\frac{1}{104}$

4. $\dfrac{3\frac{5}{7}}{6\frac{3}{4}} + \dfrac{5\frac{3}{4}}{9\frac{3}{4}} + \dfrac{2\frac{2}{3}}{6\frac{1}{2}}$

5. $\dfrac{3\frac{7}{8}}{7\frac{3}{4}} + \dfrac{4\frac{11}{12}}{14\frac{3}{4}} + \dfrac{3\frac{17}{18}}{23\frac{2}{3}}$

6. $\dfrac{1\frac{11}{13}}{3\frac{3}{7}} + \dfrac{2\frac{1}{2}}{11\frac{3}{8}} + \dfrac{17\frac{3}{7}}{40\frac{1}{5}}$

**19.**    (2) $\frac{3}{7}$ of $\frac{7\frac{5}{16}}{11\frac{1}{4}}$ $\smile$ $\frac{5}{19}$ of $3\frac{7}{16}$.

$$\frac{3}{7} \text{ of } \frac{7\frac{5}{16}}{11\frac{1}{4}} = \frac{3}{7} \times \frac{\frac{117}{16}}{\frac{45}{4}} = \frac{3}{7} \times \frac{13}{20} = \frac{39}{140} = \frac{2964}{10640}$$

$$\frac{5}{19} \text{ of } 3\frac{7}{16} = \frac{5}{19} \times \frac{55}{16} = \frac{275}{304} = \frac{9625}{10640}$$

$$\frac{9625 - 2964}{10640} = \frac{6661}{10640}.$$

We place "$\smile$" between two quantities whose difference we wish to find, when the less is written first, or when we are uncertain of their relative magnitude.

7. $\frac{5}{8}$ of $3\frac{3}{4}$ $\smile$ $\frac{11}{12}$ of $3\frac{1}{4}$

8. $\frac{19}{17}$ of $\frac{3\frac{3}{5}}{4}$ $\smile$ $\frac{2\frac{4}{5}}{5\frac{4}{7}}$

9. $\frac{11}{13}$ of $\frac{1\frac{9}{7}}{9\frac{5}{8}}$ $\smile$ $\frac{17}{42}$ of $\frac{3}{7}$

10. $\frac{7}{13}$ of $\frac{29\frac{1}{4}}{31\frac{1}{8}}$ $\smile$ $\frac{3}{11}$ of $2\frac{1}{8}$

(3) $\frac{31}{36} - (\frac{11}{20} + \frac{2}{9} - \frac{2}{3})$.

$$= \frac{155 - 99 - 40 + 120}{180} = \frac{136}{180} = \frac{34}{45}.$$

From $\frac{31}{36}$ we are required to subtract $\frac{11}{20} + \frac{2}{9}$ *diminished by* $\frac{2}{3}$. By subtracting $\frac{11}{20} + \frac{2}{9}$ we obtain a remainder too little by $\frac{2}{3}$. By adding $\frac{2}{3}$ to this remainder we therefore obtain the required result.

When "$-$" is placed before a *parenthesis*, we change the "$+$" and "$-$" signs of the enclosed quantities respectively to "$-$" and "$+$," and add or subtract as indicated by the changed signs; thus:—

$$\frac{31}{36} - (\frac{11}{20} + \frac{2}{9} - \frac{2}{3}) \text{ is} = \frac{31}{36} - \frac{11}{20} - \frac{2}{9} + \frac{2}{3}.$$

11. $\frac{7}{9} - (\frac{19}{17} - \frac{7}{13} + \frac{3}{4})$

12. $\frac{13}{19} + \frac{9}{34} - (\frac{11}{66} + \frac{4}{95})$

13. $\frac{2}{3}$ of $\frac{9}{11} - \left( \frac{3}{4} \text{ of } \frac{3}{8} - \frac{7}{9} \text{ of } \frac{3\frac{1}{3}}{5} \right)$

14. $\frac{6}{8}$ of $\frac{7}{11} - \left( \frac{17\frac{7}{11}}{37\frac{1}{5}} + \frac{3}{11} \text{ of } \frac{2\frac{3}{4}}{7} \right)$

(4) $(8\frac{3}{11} + 7\frac{23}{33}) \div \frac{7\frac{3}{8} - 5\frac{7}{9}}{\frac{2}{3} + \frac{3}{5}}$.

$$= 15\frac{32}{33} \div \frac{1\frac{8}{9}}{\frac{19}{15}} = \frac{527}{33} \div \frac{84}{57}$$

$$= \frac{527}{33} \times \frac{57}{85} = \frac{31 \times 19}{11 \times 5} = \frac{589}{55} = 10\frac{39}{55}.$$

15. $(\frac{3}{7} + \frac{5}{8}) \times (\frac{3}{8} - \frac{3}{7})$

16. $(\frac{3}{4} + \frac{5}{8} \text{ of } 3\frac{7}{8}) \times (\frac{5}{8} + \frac{3}{8} + \frac{9}{10})$

17. $(\frac{3}{8} \times \frac{19}{17}) \div (9\frac{4}{11} - 8\frac{3}{13})$

18. $\frac{\frac{3}{4} - \frac{2}{5}}{\frac{2}{3} + \frac{3}{10}} \div (\frac{3}{8} + \frac{7}{54})$

**19.** The following show the difference in value produced by changing the place of the parenthesis :—

19. $\left(\dfrac{3\frac{7}{13}}{6\frac{4}{7}} \times \frac{10}{11}\right) - \left(\dfrac{2\frac{7}{9}}{15} + \frac{4}{15} \text{ of } 1\frac{7}{18}\right).$

20. $\dfrac{3\frac{7}{13}}{6\frac{4}{7}} \times \left(\frac{10}{11} - \dfrac{2\frac{7}{9}}{15} + \frac{4}{15} \text{ of } 1\frac{7}{18}\right).$

21. $\dfrac{3\frac{7}{13}}{6\frac{4}{7}} \times \left(\frac{10}{11} - \dfrac{2\frac{7}{9}}{15}\right) + \frac{4}{15} \text{ of } 1\frac{7}{18}.$

**20.** In REDUCING the fraction of a quantity to a lower name than that in which it is given, we multiply the fraction by the number of times the former is contained in the latter; thus, in reducing $\frac{1}{27}$ *foot* to the fraction of an *inch*, we multiply the numerator by 12, and obtain $\frac{36}{27}$ *inch*, which is $= \frac{36}{27}$ of $\frac{1}{12}$ foot.

(1) Reduce $\frac{1}{27}$ oz. troy to the fraction of a grain.

$$\text{oz. } \dfrac{1 \times 20 \times \overset{8}{\cancel{24}}}{\underset{9}{\cancel{27}}} = \frac{160}{9} \text{ gr.}$$

1. $\frac{3}{58}$ s...............d.
2. $\frac{3}{80}$ £...............s.
3. $\frac{2}{83}$ gu...............s.
4. $\frac{7}{70}$ cr...............s.
5. $\frac{4}{117}$ s...............hfd.
6. $\frac{3}{35}$ cr...............sixd.
7. $\frac{2}{147}$ T...............cwt.
8. $\frac{6}{818}$ yd...............in.
9. $\frac{7}{180}$ lb. av........oz. av.
10. $\frac{1}{140}$ lb. tr.........oz. tr.
11. $\frac{11}{108}$ ac.............po.
12. $\frac{1}{33}$ da.............ho.
13. $\frac{1}{100}$ cwt.............lb.
14. $\frac{3}{100}$ sq. ft.........sq. in.
15. $\frac{3}{3520}$ ml...........yd.
16. $\frac{11}{101}$ bu.............gal.
17. $\frac{3}{880}$ fu.............yd.
18. $\frac{11}{980}$ ho.............min.

In reducing the fraction of a quantity to a higher name than that in which it is given, we divide the fraction by the number of times the former contains the latter; thus, in reducing $\frac{1}{11}$ *grain* to the fraction of a *lb. avoir.*, we multiply the denominator by 7000, and obtain $\frac{3}{77000}$ lb. avoir., which is $= \frac{3}{77000}$ of 7000 gr.

(2) Reduce $\frac{4}{7}$d. to the fraction of a crown.

$$\text{d. } \dfrac{\overset{4}{\cancel{4}}}{7} \times \dfrac{\cancel{12}}{3} \times 5 = \frac{1}{105} \text{ cr.}$$

**20.**

| | |
|---|---|
| 25. $\frac{19}{49}$ sec............ho. | 31. $\frac{4}{7}$ min.............da. |
| 26. $\frac{3}{8}$ gal. ............bu. | 32. $\frac{5}{24}$ da.............co. yr. |
| 27. $\frac{44}{7}$ yd.............ml. | 33. $\frac{18}{15}$ qt.............qr. |
| 28. $\frac{7 2}{8}$ oz. tr.........lb. tr. | 34. $\frac{27}{40}$ cub. in........cub. yd. |
| 29. $\frac{14}{9}$ pt.............gal. | 35. $\frac{22}{25}$ po.............ac. |
| 30. $\frac{28}{61}$ pk............qr. | 36. $\frac{174}{5}$ gr............lb. av. |

In reducing the fraction of a quantity to a name which is neither a measure nor a multiple of the name in which the fraction is given, we both multiply and divide as in the following example:—

(3) Reduce $\frac{73}{875}$ lb. av. to the fraction of a lb. troy.

$$\text{lb. av. } \frac{73}{875} \times \frac{\cancel{7000}^{\;8}}{\cancel{5760}_{\;720}} = \frac{73}{720} \text{ lb. tr.}$$

In multiplying by 7000, we reduce the fraction of a lb. av. to that of a grain, which, when divided by 5760, becomes that of a lb. troy.

| | |
|---|---|
| 37. $\frac{3}{4}$ fl................cr. | 42. $\frac{110}{140}$ oz. tr.........oz. av. |
| 38. $\frac{11}{4}$ gu.............£. | 43. $\frac{60}{50}$ lk..............ft. |
| 39. $\frac{7}{8}$ nl..............ft. | 44. $\frac{6}{5844}$ Ju. yr.......co. yr. |
| 40. $\frac{28}{25}$ E. E..........yd. | 45. $\frac{18}{15}$ co. mo.........co. yr. |
| 41. $\frac{41}{1780}$ lb. av......lb. tr. | 46. $\frac{120}{147}$ geog. ml.....Imp. ml. |

In reducing a compound quantity to the fraction of a simple or a compound quantity, we proceed as follows:—

(4) Reduce £1 ″ 2 ″ 7 to the fraction of £1 ″ 13 ″ 5.

$$\text{£1}\text{″}2\text{″}7 = 271d. \quad \text{£1}\text{″}13\text{″}5 = 401d.$$
$$\text{£1}\text{″}2\text{″}7 = \frac{271}{401} \text{ of £1″13″5.}$$

Having reduced the quantities to the same name, we find that since £1 ″ 2 ″ 7 contains 271 pence, of which £1 ″ 13 ″ 5 contains 401, the former is $\frac{271}{401}$ of the latter.

| | |
|---|---|
| 47. 11/6............£1 | 56. 3 oz. 4 dwt......2 lb. 6 oz. |
| 48. 2/2½............£1 | 57. 3 fu. 44 yd......3 ml. |
| 49. 2 ft. 8 in.......1 yd. | 58. 2 qr. 3 nl........3 yd. 1 qr. |
| 50. 3 ro. 15 po.....1 ac. | 59. 2 ro. 14 po......3 ac. 1 ro. |
| 51. 6 fu. 15 po.....1 ml. | 60. 7 bu. 3 pk.......1 qr. 3 bu. |
| 52. 6 oz. 3 dwt. ....1 lb. tr. | 61. 7 ho. 12 min....3 da. 4 ho. |
| 53. 4/4..............13/8 | 62. 4 da. 17 ho......1 wk. 3 da. |
| 54. 7/8¼..............13/3¾ | 63. 22° 30'..........360° |
| 55. £1″15″3.......£3″13″9 | 64. 66° 32′ 23″....90° |

**20.**   (5) Reduce $\tfrac{3}{5}$ s. to the fraction of $\tfrac{19}{27}$ £., or find what part $\tfrac{3}{5}$ s. is of $\tfrac{19}{27}$ £.

$$\overset{\text{s.}}{\tfrac{3}{5}} \times 20 = \tfrac{3}{100} = \tfrac{3 \times 27}{100 \times 10}\ \text{of}\ \tfrac{19}{27}\ \text{£.}$$
$$= \tfrac{81}{1000}\ \text{of}\ \tfrac{19}{27}\ \text{£.}$$

*Otherwise:* $\overset{\text{£.}}{\tfrac{10}{27}} \times 20 = \tfrac{200}{27}.\quad \tfrac{3}{5} \div \tfrac{200}{27} = \tfrac{3}{5} \times \tfrac{27}{200} = \tfrac{81}{1000}.$

$\tfrac{3}{5}$ s. $= \tfrac{81}{1000}$ of $\tfrac{19}{27}$ £.          *See* § 24. (2).

| | | | |
|---|---|---|---|
| 65. $\tfrac{7}{11}$ £.............$\tfrac{3}{4}$ £. | | 71. $\tfrac{4}{15}$ ac.............$\tfrac{7}{8}$ po. |
| 66. $\tfrac{4}{8}$ s.............$\tfrac{7}{8}$ s. | | 72. $\tfrac{4}{7}$ fu.............$\tfrac{7}{8}$ ml. |
| 67. $\tfrac{3}{10}$ ac.............$\tfrac{4}{8}$ ac. | | 73. $3\tfrac{1}{4}$ s.............£$2\tfrac{3}{4}$ |
| 68. $\tfrac{7}{11}$ yd.............$\tfrac{5}{12}$ yd. | | 74. $5\tfrac{7}{8}$ gal.............$1\tfrac{11}{32}$ qr. |
| 69. $\tfrac{4}{7}$ s.............$\tfrac{9}{10}$ £. | | 75. $6\tfrac{4}{5}$ ho.. .........$\tfrac{3}{4}$ da. |
| 70. $\tfrac{5}{28}$ £.............$3\tfrac{1}{7}$ s. | | 76. $\tfrac{7}{15}$ oz. av.......$\tfrac{1}{30}$ oz. tr. |

In finding the value of a fraction of a quantity, we may either work as in reducing a fraction to a lower name; or, by taking as many units of the name in which the fraction is given as are indicated by the numerator, we may divide by the denominator as in Compound Division.

(6) Find the value of $\tfrac{25}{96}$ £.

$$\text{£}\tfrac{25}{96} \times 20 = \tfrac{125}{24} = 5\tfrac{5}{24},\quad \tfrac{5}{24} \times 12 = 2\tfrac{1}{2},\quad \tfrac{25}{96} = 5 \,\text{\textit{//}}\, 2\tfrac{1}{2}.$$

*Otherwise:* £$\tfrac{25}{96} = \tfrac{1}{96}$ of £25

$$96\left\{\begin{array}{l}12 \\ 8\end{array}\right. \begin{array}{|l} \ \ \ \ \ \text{£} \\ 25 \,\text{//}\, 0 \,\text{//}\, 0 \\ \hline 2 \,\text{//}\, 1 \,\text{//}\, 8 \\ \hline 0 \,\text{//}\, 5 \,\text{//}\, 2\tfrac{1}{4} \end{array}$$

| | | |
|---|---|---|
| 77. $\tfrac{11}{24}$ s. | 84. $\tfrac{35}{128}$ lb. av. | 91. $\tfrac{11}{320}$ ml. |
| 78. $\tfrac{21}{32}$ s. | 85. $\tfrac{21}{64}$ cwt. | 92. $\tfrac{19}{32}$ ac. |
| 79. $\tfrac{113}{120}$ £. | 86. $\tfrac{19}{80}$ T. | 93. $\tfrac{17}{200}$ oz. tr. |
| 80. $\tfrac{17}{80}$ £. | 87. $\tfrac{97}{192}$ lb. tr. | 94. $\tfrac{23}{128}$ bu. |
| 81. $\tfrac{31}{48}$ cr. | 88. $\tfrac{13}{48}$ yd. | 95. $\tfrac{29}{48}$ pk. |
| 82. $\tfrac{191}{384}$ gu. | 89. $\tfrac{11}{24}$ sq. yd. | 96. $\tfrac{5}{83}$ lk. |
| 83. $\tfrac{33}{32}$ fl. | 90. $\tfrac{41}{72}$ cub. yd. | 97. $\tfrac{40}{487}$ Ju. yr. |

(7) Find the value of $\tfrac{7}{8}$ of $9\tfrac{3}{16}$ acres

$$\tfrac{7}{8}\ \text{of}\ 9\tfrac{3}{16}\ \text{ac.} = \tfrac{7}{8} \times \overset{\text{ac.}}{\tfrac{147}{16}} = \overset{\text{ac.}}{\tfrac{343}{48}} = 7\tfrac{7}{48},\quad \tfrac{7 \times 4}{48} = \tfrac{7 \times 40}{12} = 23\tfrac{1}{3},$$

$$7\tfrac{7}{48}\ \text{ac.} = 7\ \text{//}\ 0\ \text{//}\ 23\tfrac{1}{3}.$$

**20.**

Otherwise: $\overset{\text{ac.}}{\frac{3}{16}} \times \overset{}{4} = \overset{\text{ro. } 10}{\frac{3}{4}} \times \overset{\text{po.}}{40} = \overset{\text{ac.}}{30}, \ 9\frac{3}{16} \ = \ \overset{\text{ac. ro. po.}}{9 \,\text{//}\, 0 \,\text{//}\, 30}$

$$\frac{7}{9)\overline{64 \,\text{//}\, 1 \,\text{//}\, 10}}$$

$$7 \,\text{//}\, 0 \,\text{//}\, 23\tfrac{1}{3}$$

98. $\frac{3}{4}$ of $5\frac{2}{3}$ cr.  |101. $\frac{3}{5}$ of £3 // 6 // 4½  |1ა4. $\frac{4}{5}$ of $3\frac{4}{5}$ s.

99. $\frac{3}{5}$ of $3\frac{3}{4}$ hf.cr.  |102. $\frac{3}{11}$ of 3 fu. 12 po.  |105. $\frac{3}{7}$ of 2 ho.34 min.

100. $\frac{4}{5}$ of £3 // 7 // 6  |103. $\frac{4}{5}$ of 2 ac. 3 ro.  |106. $\frac{4}{5}$ of 3 wk. 6 da.

107. Express a Russian Archine, which is $\frac{7}{8}$ of a yard, as the fraction of a mile.

108. Express the height of Ben Macdhui, which is $\frac{41}{48}$ of a mile, in feet.

109. Schiehallion, where Maskelyne made a series of observations on the Density of the Earth, is nearly $\frac{5}{8}$ of a mile high.  Express its height in feet.

110. Harton Coalpit, where Airy conducted a series of observations on the Density of the Earth, is $\frac{9}{44}$ of a mile deep.  Express its depth in fathoms.

111. The velocity of sound is $\frac{7}{33}$ of a mile 𝔭 sec.  Express it in ft.

112. Express 5 dwt. 9 gr., the weight of a guinea, as the fraction of 1 lb. troy.

113. In an estate of 3173 acres 20 poles, the roads occupy 66 ac. 1 ro. 8 po.  What part of the estate is occupied by roads ?

114. The distance traversed by an express train in $\frac{1}{12}$ hour is run by a goods' train in $\frac{7}{8}$ of 1¼ hour.  What fraction is the former time of the latter ?

115. The National Subscription, promoted by Cromwell in aid of the Waldenses, amounted to £38097 // 7 // 3, of which Cromwell gave £2000.  Express the latter as the fraction of the former.

116. In November 1855, the Patriotic Fund amounted to £1,296,282 // 4 // 7, of which Glasgow subscribed £44,943 // 1 // 10.  What part was the Glasgow subscription of the whole ?

117. Express $58\frac{10}{11}$ yards, the depth of an Artesian well, as the fraction of another which is $\frac{540}{613}$ of a mile deep.

118. What fraction is an oz. avoir. of an oz. troy ?

119. Reduce a grain to the fraction of a dram avoir.

120. Express a lb. troy in avoir. weight.

121. In Scotland, during June 1856, the mean weight of vapour in a cubic foot of air was $3\frac{1}{10}$ grains.  Express this as the fraction of 1 lb. avoir.

122. In Scotland, during April 1856, the mean weight of vapour in a cubic foot of air was $\frac{48}{10000}$ lb. avoir.  Express this in grains.

123. Mont Blanc is 15780 feet above the level of the sea, and

**20.** Dhawalagiri is $5\frac{7}{80}$ miles. Express the height of the former as the fraction of that of the latter.

124. A degree of longitude on the parallel of Greenwich is nearly $= \frac{3}{5}$ of a degree of the Equator, which is $= 60 \times 6076$ ft. Find the number of Imperial miles in the former.

**21.** Find the sum of $\frac{3}{5}$ ac., $\frac{7}{8}$ of $3\frac{3}{4}$ ro., and $\frac{3}{7}$ of $16\frac{1}{8}$ po.

I.

|  | ac. | ro. | po. |
|---|---|---|---|
| $\frac{3}{5}$ ac. | $= 0$ | $2$ | $16$ |
| $\frac{7}{8}$ of $3\frac{3}{4}$ ro. | $= 0$ | $2$ | $36\frac{2}{8}$ |
| $\frac{3}{7}$ of $16\frac{1}{8}$ po. | $= 0$ | $0$ | $4\frac{2}{3}$ |
|  | $1$ | $1$ | $17\frac{1}{2}$ |

II.

| | ac. | ac. |
|---|---|---|
| $\frac{3}{5}$ ac. | $= \frac{3}{5}$ | $= \frac{144}{240}$ |
| $\frac{7}{8}$ of $3\frac{3}{4}$ ro. $= \frac{33}{12}$ ro. $=$ | $\frac{33}{48}$ | $= \frac{173}{240}$ |
| $\frac{3}{7}$ of $\frac{49}{3}$ po. $= \frac{14}{3}$ po. $=$ | $\frac{7}{240}$ | $= \frac{7}{240}$ |
| | | $\frac{324}{240}$ |

$= 1\frac{43}{120}$ ac. $= 1$ ac. $1$ ro. $17\frac{1}{2}$ po.

In adding fractions expressed in different names, we may, as in I., find the value of the fractions, and then proceed as in Compound Addition; or, as in II., we may reduce the fractions to the same name, and having added them, find the value of their sum.

1. $\frac{1}{4}$ £. $+ \frac{3}{5}$ s. $+ \frac{7}{8}$ cr.
2. $\frac{9}{10}$ £. $+ \frac{3}{4}$ fl. $+ \frac{7}{8}$ s.
3. $\frac{1}{7}$ ac. $+ 2\frac{11}{14}$ ro. $+ 3\frac{1}{2}$ po.
4. $\frac{2}{3}$ ml. $+ \frac{3}{13}$ fu. $+ \frac{7}{11}$ po.
5. $\frac{3}{4}$ lb. $+ 1\frac{2}{7}$ oz. $+ 2\frac{2}{3}$ dwt.
6. $1\frac{43}{150}$ £. $+ \frac{2}{5}$ s. $+ 1\frac{1}{4}$ of $\frac{7}{22}$ £.

7. $\frac{4}{5}$ T. $+ \frac{3}{4}$ cwt. $+ \frac{6}{7}$ qr.
8. $\frac{3}{14}$ qr. $+ \frac{7}{8}$ bu. $+ \frac{3}{4}$ pk. $+ \frac{5}{8}$ gal.
9. $\frac{7}{8}$ T. $+ 7\frac{3}{4}$ cwt. $+ 1\frac{13}{16}$ qr. $+ 20\frac{2}{3}$ lb.
10. $\frac{7}{50}$ da. $+ \frac{4}{5}$ ho. $+ 6\frac{3}{4}$ min. $+ \frac{9}{34}$ ho.
11. $\frac{7}{18}$ ft. $+ \frac{3}{4}$ yd. $+ \frac{7}{110}$ fu.
12. $\frac{3}{8}$ of $3\frac{4}{8}$ po. $+ \frac{7}{80}$ ml. $+ \frac{3}{7}$ of $2\frac{5}{8}$ fu.

13. Find the total weight of seven half-chests of tea, containing respectively $1\frac{4}{7}$ qr., $\frac{7}{13}$ cwt., $\frac{8}{217}$ T., $1\frac{4}{7}$ qr., $\frac{13}{14}$ cwt., $\frac{1}{4}$ of $\frac{6}{15}$ T., and $\frac{7}{8}$ cwt.

14. How many acres are in a parish in which cultivated land occupies $2\frac{13}{16}$ sq. miles; pasture, $\frac{2}{7}$ of $13\frac{7}{18}$ sq. miles; and plantation, $234\frac{3}{4}$ acres?

15. Find the weight, by the old system, of a pill-mass, consisting of 1 ʒ rhubarb, $\frac{1}{2}$ ʒ acetate of potash, and $1\frac{1}{4}$ ʒ of conserve of roses.

16. The highest part of the woody region of Mount Etna is $\frac{3\frac{1}{3}}{4}$ of $1\frac{1\frac{1}{3}}{4}$ mile above the level of the sea; the foot of the cone is $1160\frac{1}{2}$ yd. higher; and the summit is $1\frac{1}{3}$ of $1316\frac{6}{11}$ ft. above the latter. Find the height of the summit above the level of the sea?

**22.** (1) From $\frac{3}{5}$ of $6\frac{1}{8}$ fur. subtract $\frac{3}{14}$ mile; or find the value of $\frac{3}{5}$ of $6\frac{1}{8}$ fu. $- \frac{3}{14}$ ml.

I

| | fu. | po. | yd. |
|---|---|---|---|
| $\frac{3}{5}$ of $6\frac{1}{8}$ fu. | $= 3$ | $32$ | $0$ |
| $\frac{3}{14}$ ml. | $= 1$ | $28$ | $3\frac{1}{7}$ |
| | $2$ | $3$ | $2\frac{6}{14}$ |

II.

| | fu. |
|---|---|
| $\frac{3}{5}$ of $6\frac{1}{8}$ fu. | $= 3\frac{4}{5}$ | $= 3\frac{11}{15}$ |
| $\frac{3}{14}$ ml. | $= 1\frac{5}{7}$ | $= 1\frac{11}{15}$ |

2 fu. 3 po. $2\frac{6}{14}$ yd. $= 2\frac{3}{15}$ fu.

**22.** In finding the difference between fractions expressed in different names, we may, as in I., find the value of the fractions, and then proceed as in Compound Subtraction; or, as in II., we may reduce the fractions to the same name, and find the value of their difference.

In I. we have the number of yards $= 5\frac{1}{2} - 3\frac{1}{7} = 2 + \frac{7-2}{14} = 2\frac{5}{14}$.

1. $\frac{5}{8}$ £. — $\frac{7}{8}$ s.
2. $\frac{11}{15}$ cwt. — $\frac{14}{15}$ qr.
3. $\frac{2}{15}$ cr. ⌣ $\frac{3}{5}$ £.
4. $\frac{7}{8}$ cwt. — $\frac{3}{70}$ T.
5. $\frac{2}{3}$ oz. — $\frac{1}{15}$ dwt.
6. $\frac{11}{40}$ pk. ⌣ $\frac{5}{8}$ bu.

(2) Find the value of $\frac{4}{5}$ £. — ($\frac{7}{8}$ s. + $\frac{3}{10}$ cr. — $\frac{3}{5}$ fl.)

$\frac{4}{5}$ £. — ($\frac{7}{8}$ s. + $\frac{3}{10}$ cr. — $\frac{3}{5}$ fl.) = $\frac{4}{5}$ £. — $\frac{7}{8}$ s. — $\frac{3}{10}$ cr. + $\frac{3}{5}$ fl.

|   | s. | d. |
|---|----|----|
| $\frac{4}{5}$ £. = | 16 ″ | 0 |
| $\frac{3}{5}$ fl. = | 1 ″ | $2\frac{1}{4}\,\frac{3}{5}$ |
|   | 17 ″ | $2\frac{1}{4}\,\frac{3}{5}$ |

|   | s. | d. |
|---|----|----|
| $\frac{7}{8}$ s. = | 0 ″ | $10\frac{1}{2}$ |
| $\frac{3}{10}$ cr. = | 1 ″ | 6 |
|   | 2 ″ | $4\frac{1}{2}$ |

$17 \text{″} 2\frac{1}{4}\,\frac{3}{5} - 2 \text{″} 4\frac{1}{2} = 14 \text{″} 9\frac{3}{4}\,\frac{3}{5}$.

7. $\frac{3}{4}$ £. + $\frac{7}{18}$ s. — ($\frac{7}{20}$ cr. — $\frac{3}{8}$ fl. + $\frac{5}{8}$ gu.)

8. $\frac{9}{50}$ ac. — ($\frac{7}{8}$ ro. + $\frac{13}{18}$ po. — $\frac{5}{14}$ ac.)

9. By how much does $1\frac{1}{13}$ jacobus exceed $\frac{3}{4}$ joannes?

10. A vessel containing $\frac{5}{8}$ gal. is filled, and $\frac{1}{4}$ of $3\frac{1}{2}$ pt. is then poured out. How much is left in the vessel?

11. An apothecary prepares $\frac{9}{14}$ ℥ of medicine, which contains $1℈$ 4 gr. of conserve of roses. What is the weight of the other ingredients, by the old system?

12. The rope of a bucket, while ascending the shaft of a coal pit $\frac{1}{2}$ of $212\frac{3}{4}$ fathoms deep, snaps while the bucket is $\frac{1}{3}$ of $200\frac{3}{4}$ ft. from the top. Through what depth is the bucket precipitated?

13. The top of St Peter's, Rome, is $\frac{13}{110}$ mile above the ground, while that of St Paul's, London, is $\frac{17}{304}$ mile. Express their difference in feet.

14. A retail grocer having bought $\frac{6}{7}$ of $58\frac{1}{4}$ lb. of tea, sold during six days, $\frac{1}{3}$ qr. $\frac{8}{12}$ cwt., $\frac{1}{4}$ qr. $\frac{7}{18}$ cwt., $\frac{3}{8}$ qr., and $\frac{3}{4}$ of $\frac{7}{17}$ of $18\frac{3}{8}$ lb. How many lb. has he still on hand?

15. A draper having a piece of cloth containing $27\frac{1}{4}$ yd., sells $\frac{5}{8}$ of $7\frac{1}{14}$ yd., $\frac{1}{2}$ of $3\frac{3}{4}$ yd., and $\frac{1}{4}$ of 3 qr. What has he over?

16. The astronomical stations chosen by Professor Piazzi Smyth in Teneriffe, in 1856, were respectively $\frac{1}{2}\frac{9}{33}$ of $2\frac{1}{2}$ miles and $\frac{4}{7}\frac{9}{9}$ of $3\frac{440}{1920}$ miles above the level of the sea. By how many yards did the height of the latter exceed that of the former?

**23.** Multiply $\frac{3}{7}$ £. by $30\frac{3}{4}$.

$30\frac{3}{4} \times \frac{3}{7}$ £. $= \frac{123}{4} \times \frac{3}{7}$ £. $= £13\frac{5}{28} = £13''3''6\frac{3}{4}\frac{3}{7}$.

We multiply the fraction of a quantity as in abstract numbers, and then find the value of the product.

1. $\frac{4}{5}$ £ $\times$ 17
2. $\frac{3}{8}$ s. $\times$ $29\frac{3}{4}$
3. $\frac{3}{11}$ ac. $\times$ 18
4. $1\frac{5}{8}$ pk. $\times$ $3\frac{11}{13}$
5. $\frac{7}{50}$ ho. $\times$ $\frac{21}{49}$
6. $\frac{9}{50}$ ml. $\times$ $\frac{7}{9\frac{1}{2}}$

7. An incumbent has received 40 stipends at an average of £148$\frac{1}{2}\frac{9}{8}$ each. Find the total amount.

8. If a train runs a mile in $\frac{9}{50}$ hour, in what time will it traverse $\frac{3}{4}$ of 150 miles?

9. Find the price of $6\frac{3}{4}$ pieces, each $29\frac{3}{4}$ yd., @ $2\frac{7}{4}$s. ℈ yd.

10. A farmer having found 263 sheep trespassing on his fields, claims by an old statute, as compensation from their owner, $\frac{1}{2}$ of $\frac{1}{4}$ of £$\frac{2}{12}$ for each sheep. Find the total claim.

11. A train runs $\frac{3}{4}$ mile in a minute; what distance will it run in $\frac{1}{4}$ of $3\frac{3}{4}$ hours?

12. The area of Paris is $6\frac{7}{11}\frac{7}{8}$ times as large as that of Frankfort-on-Maine. which is$=2\frac{1}{10}$ sq. miles. Express the former in acres.

13. The area of one of the parishes in the smallest county in Great Britain is $\frac{1}{4}\frac{5}{9}\frac{8}{8}$ of 4563 acres, while that of the county is $6\frac{1}{8}\frac{2}{3}\frac{5}{7}$ times as large. Express the area of the latter in sq. miles.

14. The $\frac{3}{4}$ of a Prussian thaler is pure silver. The weight of a thaler is $\frac{9}{11}$ of a Cologne mark, which is $= 7\frac{1}{16}$ oz. troy. How much pure silver is in a thaler?

**24.** (1) Divide $1\frac{7}{11}$ acre by $28\frac{4}{7}$.

$1\frac{7}{11}$ ac. $\div$ $28\frac{4}{7} = \frac{18}{11}$ ac. $\times \frac{7}{400} = \frac{133}{2400}$ ac. $= 8\frac{13}{15}$ po.

We divide the fraction of a quantity as in abstract numbers, and then find the value of the quotient.

(2) How often is $\frac{4}{9}$ s. contained in $\frac{4}{11}$ £.?

$\frac{4}{9}$ s. $= \frac{1}{45}$ £.

$\frac{4}{11}$ £. $\div \frac{1}{45}$ £. $= 10\frac{2}{11}$ times.

In dividing one quantity by another, we reduce them both to the same name, and by finding the quotient, we see how many times the one is contained in the other.

This operation is equivalent to finding the fraction, proper or improper, which the dividend is of the divisor; thus, as in § 20. (5), we find that $\frac{4}{11}$ £. is $= \frac{112}{11}$ of $\frac{4}{9}$ s., or that $\frac{4}{11}$ £. is $= 10\frac{2}{11}$ times $\frac{4}{9}$ s.

1. $13\frac{7}{18}$ s. $\div$ $20\frac{1}{8}$
2. $\frac{4}{7}$ cr. $\div$ $1\frac{4}{18}$
3. $1\frac{7}{18}$ ml. $\div$ $5\frac{1}{9}$
4. $8\frac{1}{3}$ da. $\div$ $1\frac{8}{37}$
5. $9\frac{11}{14}$ ac. $\div$ $\frac{17}{24}$ ac.
6. $4\frac{7}{33}$ sq. yd. $\div$ $\frac{13}{44}$ sq. yd.

**24.**
7. 623 sovereigns are coined out of $\frac{3}{8}$ of 19$\frac{1}{2}$ lb. troy of sterling gold. Find the weight of a sovereign.

8. 155 Napoleon pieces weigh 32$\frac{37}{248}$ oz. troy. Find the weight of a Napoleon piece.

9. If a cubic foot of air contains 2$\frac{1}{30}$ grains of vapour; what volume of air will contain 1 lb. avoir. of vapour?

10. How many crofts, each $\frac{43}{51}$ of 3$\frac{1}{2}$ roods, can be portioned out of 121 acres?

11. How many pieces, each $\frac{41}{64}$ of 48 yards, are contained in $\frac{9}{50}$ of 683$\frac{1}{4}$ E. E.?

12. How many lb. troy, each $\frac{141}{242}$ lb. av., are $=\frac{5}{8}$ of $\frac{31}{4}$ cwt.?

13. In Mid-Lothian, the total area under a rotation of crops was, in 1856, 104077$\frac{3}{4}$ acres, and in 1857, 160$\frac{9}{11}\frac{9}{2}$ square miles. What part of the former is the latter?

14. An American dollar weighs $\frac{5}{6}\frac{4}{5}$ oz. troy, and a British crown $\frac{5}{11}$ lb. troy. Express the former as the fraction of the latter.

**25.**    MISCELLANEOUS EXERCISES IN VULGAR FRACTIONS.

1. How many hundredths of an inch are in a link?

2. A student has read $\frac{5}{7}$ of the Sixth Book of the Æneid, which contains 903 lines. How many lines has he yet to read?

3. Find the weight of 200 guineas, each 5$\frac{1}{8}$ dwt.

4. The sheriff and justices of peace of a county enrolled 54 special constables in one day, on the next day $\frac{2}{3}$ of that number, and on the third day $\frac{3}{4}$ of the number enrolled on the second. How many have been enrolled in all?

5. A boy who has 36 marbles gains $\frac{1}{4}$ of that number, and then loses $\frac{7}{12}$ of what he has. How many marbles has he gained?

6. In 1855, the population of Texas, amounting to 400,000, included 35,000 Germans. What part of the entire population was the rest of the inhabitants?

7. In 1856, 106000 acres in Ireland were occupied in the growth of flax, of which 150 square miles were in Ulster. What part is the latter of the whole?

8. Of a vessel, worth £5600, A, who has $\frac{11}{64}$, sells $\frac{3}{4}$ of his share to B, who sells $\frac{2}{3}$ of his to C. Find the value of C's share.

9. Of a number of sheep on a hill-farm, the Cheviot ewes were $\frac{1}{4}$, the black-faced ewes $\frac{2}{9}$, the Cheviot hogs $\frac{18}{53}$, the half-bred hogs $\frac{16}{53}$, and the remainder consisted of 100 black-faced hogs. Find the total number.

10. In Scotland, in 1855, the number of deaths in February, the month of greatest mortality in that year, was 7227; and in September, the month of least mortality in 1855, the number of deaths

**25.** was 32 more than $\frac{2}{3}$ of that in February. Find the number in September.

11. If, in small farms in Asia Minor, $\frac{1}{2}$ of the produce is given to the landlord who furnishes the seed, and $\frac{1}{15}$ of the remainder to the government as land-tax, what part remains to the tenant?

12. A gentleman leaves property worth £556 to his cousin, who pays a duty amounting to $\frac{3}{10}$ of its value; and £470 to his second cousin, who pays $\frac{3}{10}$ of it in duty. Find the total duty on both.

13. A bankrupt's effects amount to $\frac{2}{3}$ of $\frac{3}{4}$ of his debts. How much can he pay per £.?

14. A bankrupt pays 11/3 $\frac{1}{p}$ £. What part of his debts are his effects?

15. In the examination for admission to the Royal Military Academy at Woolwich, the number of marks for English amounts to 1250, and is $\frac{3}{4}$ of the number of marks for Mathematics. Find the number of the latter.

16. Divide £57$\frac{1}{2}$ into 4$\frac{1}{2}$ shares.

17. Divide £819 among 6 men and 5 youths, giving a youth $\frac{2}{3}$ of a man's share.

18. Share a bonus of £20$\frac{1}{2}$ among 1 foreman, 16 journeymen, and 4 apprentices, giving a journeyman $\frac{1}{3}$ of the foreman's share, and an apprentice $\frac{1}{15}$ of a journeyman's.

19. Sir George Cathcart, who fell at Inkerman in 1854, was 16 when he received his commission. He spent $\frac{1}{4}$ of his life in the military profession. In what year was he born?

20. In the end of 1855, the number of widows relieved by the Patriotic Fund, amounting to 2544, was $\frac{1}{3}$ of that of children relieved. Find the number of the children.

21. The copper sheathing of the hull of a vessel which had been seven years in the Pacific was found to contain $\frac{1}{100}$ of its weight in silver. What fraction of a lb. troy of silver would 1 cwt. of the sheathing contain?

22. In the division in the House of Commons on March 3, 1857, on the Canton disturbances, among those who voted against the Ministry there were 198 Conservatives, and the numbers of Peelites and Liberals were respectively $\frac{1}{3}$ and $\frac{5}{9}$ of this number; while of those who voted with the Ministry the number of Liberals was $5\frac{1}{2}$ times that of Liberals on the other side, and the number of Conservatives $\frac{7}{8}$ of that of the opposite Conservatives. Find the majority against the Ministry.

23. In 1856, the number of births in the eight principal towns of Scotland was 31885. Find the number of deaths, which was 527 less than $\frac{2}{3}$ of that of births.

C

**25.**

24. Montaigne the Essayist's copy of Cæsar's Commentaries was bought at a bookstall for $\frac{9}{10}$ franc, and subsequently sold by auction for 1550 francs. How many times does the latter contain the former?

25. From Montreal to Toronto by the Grand Trunk Railway is 332 miles. Of this, $\frac{1}{2}$ mile more than $\frac{5}{8}$ was opened in November 1855, and the remainder in November 1856. Find the latter distance.

26. The 36 Israelites who fell in the first assault on Ai were $2\frac{8}{10}$ of the force sent by Joshua. How many were there in all?

27. Of 909 men of the 23d Foot or Royal Welsh Fusiliers, 32 men more than $\frac{5}{8}$ were killed and wounded in the Crimea. How many were killed and wounded?

28. In the Line, the price of a lieutenant-colonel's commission is £4500, a major's is $\frac{11}{15}$ of a lieut.-colonel's, a captain's $\frac{9}{10}$ of a major's, a lieutenant's $\frac{7}{8}$ of a captain's, and an ensign's $\frac{9}{14}$ of a lieutenant's. Find the price of an ensign's commission.

29. Of 98600 non-commissioned officers and privates in the British service who sailed for the Crimea, 25500 embarked under Lord Raglan. What fraction was the remainder of the whole?

30. An angler for fishing salmon smolts was fined £$1\frac{7}{10}$. The expenses of court were $2\frac{3}{8}$ times the fine. Find the whole amount.

31. Of the number in the British Army killed and wounded in the Crimea until the fall of Sebastopol, in siege-duties there were 54 men more than $\frac{13}{48}$, in assaults 115 fewer than $\frac{13}{48}$, and in battles 408 more than $\frac{13}{48}$. Find the total number.

32. Of the number of shares in the Atlantic Telegraph Company, 4 shares more than $\frac{6}{25}$ are held in America, 1 more than $\frac{2}{9}$ in London, 16 more than $\frac{1}{6}$ in Liverpool, 2 more than $\frac{1}{10}$ in Glasgow, $\frac{2}{15}$ in Manchester, and $\frac{1}{35}$ in other places in Great Britain. Find the total number of shares.

33. A alone can do a work in $6\frac{1}{2}$ days, and with B's assistance in $3\frac{1}{10}$ days. In what time will B do it by himself?

34. What number multiplied by $8\frac{2}{3}$ is $=3\frac{1}{2}+\frac{3}{4}+\frac{1}{2}\frac{6}{3}+\frac{4}{3}\frac{3}{10}$?

35. Multiply the sum of $\frac{3}{4}$, $\frac{4}{5}$, and $\frac{5}{6}$ by the difference between $\frac{4}{5}$ and $\frac{4}{7}$, and divide the product by the sum of $\frac{4}{5}$ and $\frac{5}{8}$.

36. Multiply the sum of $\frac{9}{10}$ and $\frac{3}{4}$ by their difference.

37. Find that number, to which, if we add $\frac{9}{10}$ of $6\frac{7}{8}$, the result will be $\frac{5}{8}$ of $13\frac{1}{2}$.

38. What number when multiplied by $\frac{3}{8}$ of $5\frac{1}{3}$ gives the product $16\frac{4}{7}$?

39. Multiply the product of $1\frac{7}{13}$ and $\frac{2\frac{2}{3}}{15}$ by the quotient of the former by the latter.

**25.**  40. There were 154 fewer wrecks on the coasts of the United Kingdom in 1855 than in 1854, and this difference was $\frac{22}{111}$ of the number in 1854. Find the number of wrecks in 1855.

41. Find the content of a plank $23\frac{3}{4}$ ft. long and $5\frac{7}{8}$ in. broad.

42. How many square feet are in a wall $5\frac{7}{8}$ yd. long and $6\frac{1}{4}$ ft. high?

43. What is the circumference of a room whose opposite walls are equal, the length being $30\frac{1}{4}$ ft. and the breadth $22\frac{7}{11}$ ft.

44. How many square yards are in the walls of a room $26\frac{1}{2}$ ft. long, $18\frac{3}{4}$ ft. broad, and $14\frac{9}{11}$ ft. high?

45. How many cubic ft. are in a box $5\frac{4}{5}$ ft. long, $2\frac{3}{5}$ ft. broad, and $23\frac{1}{4}$ in. deep?

46. A can do a work in $\frac{6}{7}$ of the time which B can, and C can do it in $\frac{11}{13}$ of A's time. They take $10\frac{1}{2}$ days, working together. In what time can each do it?

47. A cistern can be filled by a pipe in $14\frac{3}{4}$ minutes, and emptied by another in 18 minutes. In what time will it be filled when both the pipes are open?

48. In a map drawn on the lineal scale of $\frac{1}{2400}$ of that of nature, how many inches represent a mile?

49. The height of Kinchin-junga in the Himalayas, above the level of the sea, is $5\frac{9.9}{11}$ miles, and that of Aconcagua in the Andes is 150 feet greater than $4\frac{1}{4}$ miles. Reduce the latter to the fraction of the former.

50. The attraction of gravity at the Equator is less than that at the Poles by $\frac{1}{185}$ on account of centrifugal force, and $\frac{1}{515}$ on account of the earth's oblateness. Find the sum of these fractions, and give a fraction with the numerator 1, to which the sum is nearly equal.

---

# DECIMAL FRACTIONS.

In Integers we employ the decimal notation, by which the places ascending from right to left have respectively the local value of *units, tens, hundreds, thousands,* &c. Fractions in which the decimal notation is employed are termed DEC-IMAL FRACTIONS. In Decimal Fractions, the places descending from left to right have respectively the local value of *tenths, hundredths, thousandths,* &c. Thus, in 4·235, the point is placed to the right of the units' place, and the figures to the right of the point represent 2 tenths, 3 hundredths, 5 thousandths; ·235 denotes $\frac{2}{10} + \frac{3}{100} + \frac{5}{1000} = \frac{200 + 30 + 5}{1000} = \frac{235}{1000}$; and 4·235 $= 4\frac{235}{1000}$. Similarly, ·0379 denotes $\frac{3}{100} + \frac{7}{1000} + \frac{9}{10000} = \frac{300 + 70 + 9}{10000} = \frac{379}{10000}$.

**26.** A Decimal Fraction may be expressed in the form of a vulgar fraction, having the figures of the decimal as the numerator, and 10, or a power of 10, as 100, 1000, &c., as the denominator. The number of figures in the decimal is = the number of ciphers annexed to " 1 " in the denominator of the vulgar fraction.

Ciphers annexed to a decimal do not alter its value; thus, $\cdot36 = \cdot360 = \cdot3600$, for $\frac{36}{100} = \frac{360}{1000} = \frac{3600}{10000}$.

Express the following decimals in the form of vulgar fractions:—

(1) $\cdot1341 = \frac{1341}{10000}$.    (2) $\cdot00739 = \frac{739}{100000}$.

| | | |
|---|---|---|
| 1. $\cdot3$ | 5. $\cdot4153$ | 9. $\cdot009$ |
| 2. $\cdot27$ | 6. $\cdot8827$ | 10. $\cdot0007$ |
| 3. $\cdot167$ | 7. $\cdot32471$ | 11. $\cdot000093$ |
| 4. $\cdot231$ | 8. $\cdot98347$ | 12. $\cdot000107$ |

(3) $\cdot005 = \frac{5}{1000} = \frac{1}{200}$.    (4) $\cdot0848 = \frac{848}{10000} = \frac{53}{625}$.

| | | |
|---|---|---|
| 13. $\cdot8$ | 17. $\cdot032$ | 21. $\cdot0425$ |
| 14. $\cdot125$ | 18. $\cdot004$ | 22. $\cdot46875$ |
| 15. $\cdot3125$ | 19. $\cdot0625$ | 23. $\cdot00256$ |
| 16. $\cdot15625$ | 20. $\cdot7168$ | 24. $\cdot000375$ |

Write the following fractions in the form of decimals:—

(5) $\frac{71}{100} = \cdot71$.    (6) $\frac{3}{1000} = \cdot003$.

| | | |
|---|---|---|
| 25. $\frac{7}{10}$ | 29. $\frac{217}{10000}$ | 33. $\frac{3219}{100000}$ |
| 26. $\frac{8}{100}$ | 30. $\frac{83}{1000}$ | 34. $\frac{14159}{100000}$ |
| 27. $\frac{17}{1000}$ | 31. $\frac{17}{10000}$ | 35. $\frac{287}{10000}$ |
| 28. $\frac{213}{1000}$ | 32. $\frac{9}{100000}$ | 36. $\frac{307}{1000000}$ |

**27.** By moving the decimal point of a number one place towards the *right*, we *increase* the value of the number tenfold; thus, $\cdot34 \times 10 = 3\cdot4$; $\cdot07 \times 10 = \cdot7$. By moving the decimal point of a number one place towards the *left*, we *diminish* the value of the number tenfold; thus, $7\cdot13 \div 10 = \cdot713$; $\cdot79 \div 10 = \cdot079$.

In *multiplying* a decimal by a power of 10, we move the point as many places towards the *right* as there are ciphers in the multiplier; and in *dividing* by a power of 10, we move it as many places towards the *left* as there are ciphers in the divisor.

(1) Multiply and Divide $\cdot00347$ by 1000.

$$\cdot00347 \times 1000 = 3\cdot47$$
$$\cdot00347 \div 1000 = \cdot00000347.$$

**27.**
(2) Multiply 3·219 by 10000.
$$3\cdot219 \times 10000 = 32190.$$
(3) Divide 7830 by 100000.
$$7830 \div 100000 = \cdot0783.$$

| | |
|---|---|
| 1. $\cdot0369 \times 1000$ | 7. $\cdot273 \div 100$ |
| 2. $\cdot2176 \times 100$ | 8. $\cdot5236 \div 1000$ |
| 3. $\cdot42839 \times 10000$ | 9. $\cdot367 \div 10000$ |
| 4. $3\cdot216 \times 1000$ | 10. $72\cdot3 \div 100$ |
| 5. $7\cdot23 \times 10000$ | 11. $98\cdot475 \div 1000$ |
| 6. $15\cdot9 \times 10000$ | 12. $8\cdot375 \div 10000$ |

**28.** To reduce a vulgar fraction, as $\frac{3}{8}$, to a decimal, we must multiply the numerator and the denominator by such a number as will produce a power of 10 in the denominator.

Since 1000 is the lowest power of 10 which contains 8, we multiply the numerator and the denominator of $\frac{3}{8}$ by $\frac{1000}{8}$, which is $= 125$. $\frac{3}{8} = \frac{3 \times 125}{8 \times 125} = \frac{375}{1000} = \cdot375$. Now, $3 \times 125 = 3 \times \frac{1000}{8} = \frac{3000}{8}$; hence the figures of the decimal are obtained by annexing ciphers to the numerator of the vulgar fraction and dividing by the denominator. The number of places in the decimal is $=$ the number of annexed ciphers.

When we can readily find how often the lowest power of 10, which is a multiple of the denominator, contains it, we may multiply the numerator by the quotient; thus, $\frac{17}{1250} = \frac{17 \times 8}{10000} = \frac{136}{10000} = \cdot0136.$

Since the prime factors of 10 are 2 and 5, no number containing any other prime factor will exactly divide a power of 10. Hence, those Vulgar Fractions only whose denominators in the lowest terms of the fraction have no other prime factor than 2 or 5, produce TERMINATE DECIMALS.

Express the following vulgar fractions as decimals :—

(1) $\frac{3}{4} = \cdot75.$

$\begin{array}{r} 4)3\cdot00 \\ \hline \cdot75 \end{array}$ or $\frac{3}{4} = \frac{75}{100} = \cdot75.$

(2) $\frac{7}{125} = \cdot056.$

$\begin{array}{r} 125)7\cdot000 \\ \hline \cdot056 \end{array}$ or $\frac{7}{125} = \frac{56}{1000} = \cdot056.$

(3) $\frac{3}{400} = \frac{1}{100}$ of $\frac{3}{4} = \cdot0075.$

| | | | | |
|---|---|---|---|---|
| 1. $\frac{1}{8}$ | 7. $\frac{4}{25}$ | 13. $\frac{9}{64}$ | 19. $\frac{13}{500}$ | 25. $\frac{31}{320}$ |
| 2. $\frac{1}{4}$ | 8. $\frac{13}{25}$ | 14. $\frac{7}{80}$ | 20. $\frac{17}{640}$ | 26. $\frac{183}{625}$ |
| 3. $\frac{1}{6}$ | 9. $\frac{7}{8}$ | 15. $\frac{3}{125}$ | 21. $\frac{23}{160}$ | 27. $\frac{222}{500}$ |
| 4. $\frac{3}{5}$ | 10. $\frac{5}{16}$ | 16. $\frac{7}{128}$ | 22. $\frac{7}{800}$ | 28. $\frac{233}{256}$ |
| 5. $\frac{5}{8}$ | 11. $\frac{11}{32}$ | 17. $\frac{31}{160}$ | 23. $\frac{11}{1600}$ | 29. $\frac{1}{512}$ |
| 6. $\frac{7}{8}$ | 12. $\frac{3}{40}$ | 18. $\frac{11}{250}$ | 24. $\frac{19}{64}$ | 30. $\frac{611}{?}$ |

**29.** In the ADDITION of Decimals, we place tenths under tenths, hundredths under hundredths, &c., and thus add figures having the same local value.

   (1) 67·37 + ·1883 + ·0965 + 6·314 + 77·4006.

We carry as in integers; thus, for 14 ten thousands, we write 4 in the ten thousandths' place, and carry 1 to the thousandths' column. Similarly with the thousandths and the hundredths. For 13 tenths, we write 3 in the tenths' place, and carry 1 to the units' column.

$$
\begin{array}{r}
67{\cdot}37 \\
{\cdot}1883 \\
{\cdot}0965 \\
6{\cdot}314 \\
77{\cdot}4006 \\
\hline
151{\cdot}3694
\end{array}
$$

1. ·30103 + ·47712 + ·60206 + ·69897
2. ·096 + ·0096 + 96·0096 + ·96
3. 7·0096 + ·314 + ·326 + 81·093 + 325·73
4. ·7146 + ·003 + 94·216 + ·314 + 95·279
5. 93·423 + ·875 + ·329 + 4·326 + 57·916
6. 373·912 + 37·3912 + 3739·12 + 3·73912
7. 247·35 + 9·168 + ·709 + 82·361 + 18·017
8. ·73 + ·0073 + ·073 + ·00073 + ·000073
9. ·716 + ·00716 + 716·0716 + ·0000716

   (2) Add $\frac{3}{4}$, $\frac{7}{8}$, and $\frac{5}{16}$ by Vulgar and Decimal Fractions.

$$
\begin{array}{rcccl}
\frac{12}{16} &=& \frac{3}{4} &=& \cdot75 \\
\frac{14}{16} &=& \frac{7}{8} &=& \cdot875 \\
\frac{5}{16} &=& \frac{5}{16} &=& \cdot3125 \\
\hline
\frac{31}{16} &=& 1\frac{15}{16} &=& 1{\cdot}9375
\end{array}
$$

10. $\frac{3}{4}+\frac{5}{8}+\frac{3}{16}+\frac{9}{16}$
11. $\frac{7}{8}+\frac{4}{5}+\frac{3}{8}+\frac{7}{16}$
12. $\frac{3}{5}+\frac{7}{25}+\frac{11}{50}+\frac{9}{100}$
13. $\frac{8}{25}+\frac{13}{100}+\frac{17}{250}+\frac{111}{1250}$
14. $\frac{3}{8}+\frac{7}{16}+\frac{3}{32}+\frac{104}{125}+\frac{7}{625}$
15. $\frac{4}{5}+\frac{9}{32}+1\frac{3}{25}+\frac{63}{80}$

**30.** In the SUBTRACTION of Decimals, we find the difference between figures of the same local value.

   (1) ·59 — ·043.

By taking 3 thousandths from 10 thousandths, we obtain 7, which we write in the thousandths' place. We proceed as in integers, taking 5 from 9, or 4 from 8, &c.

$$
\begin{array}{r}
{\cdot}59 \\
{\cdot}043 \\
\hline
{\cdot}547
\end{array}
$$

1. ·5475 — ·4212
2. ·875 — ·525
3. ·275 — ·198
4. 5·25 — 3·875
5. 3·125 — 1·9375
6. 8·425 — 5·3875
7. 1·25 — ·175
8. 2·834 — 2·786
9. 3·245 — 1·2375
10. 1·1 — ·0009
11. 8·75 — 7·00009
12. 9·03 — ·90003

**30.**  (2) Subtract $\frac{7}{16}$ from $\frac{11}{16}$ by Vulgar and Decimal Fractions.

$$\tfrac{176}{400} = \tfrac{11}{25} = \cdot 44$$
$$\tfrac{174}{400} = \tfrac{7}{16} = \cdot 4375$$
$$\overline{\qquad \tfrac{1}{400} \qquad = \qquad \cdot 0025}$$

13. $\frac{3}{4} - \frac{3}{8}$  |  15. $\frac{3}{5} - \frac{7}{16}$  |  17. $\frac{19}{35} - \frac{14}{25}$
14. $3\frac{1}{3} - \frac{7}{8}$  |  16. $\frac{7}{8} - \frac{4}{5}$  |  18. $\frac{17}{40} - \frac{3}{8}$

**31.**  In the MULTIPLICATION of Decimals we proceed as in integers, and point off as many decimal places in the product as there are together in the multiplicand and the multiplier.

(1) Multiply ·347 by 2·3.

$$\cdot 347 \times 2\cdot 3 = \tfrac{347}{1000} \times \tfrac{23}{10}$$
$$= \tfrac{7981}{10000} = \cdot 7981.$$

$$\begin{array}{r} \cdot 347 \\ 2\cdot 3 \\ \hline 1041 \\ 694 \\ \hline \cdot 7981 \end{array}$$

In working by vulgar fractions, we see that the number of ciphers in the denominator of the product is = the sum of the numbers of the ciphers in the denominators of the factors; so, the number of decimal places in the product is = the sum of the numbers of decimal places in the factors.

(2) Multiply ·53 by ·0047.

$$\cdot 53 \times \cdot 0047 = \tfrac{53}{100} \times \tfrac{47}{10000}$$
$$= \tfrac{2491}{1000000} = \cdot 002491.$$

$$\begin{array}{r} \cdot 53 \\ \cdot 0047 \\ \hline 371 \\ 212 \\ \hline \cdot 002491 \end{array}$$

(3) Multiply ·74213 by 700.

Since one factor contains *five* decimal places, and the other ends in *two* ciphers, we point off *three* places in the product.

$$\begin{array}{r} \cdot 74213 \\ 700 \\ \hline 519\cdot 491 \end{array}$$

(4) Multiply 5·09 by 67000.

Since one factor contains *two* decimal places, and the other ends in *three* ciphers, we annex *one* cipher to the product.

$$\begin{array}{r} 5\cdot 09 \\ 67000 \\ \hline 3563 \\ 3054 \\ \hline 341030 \end{array}$$

1. 5·27×4·83  |  7. 5·27×·00483  |  13. 52·7×48300
2. ·436×2·19  |  8. ·0436×·00219  |  14. 4·36×219000
3. 1·89×·76  |  9. 18·9×·000076  |  15. ·189×7600
4. 2·38×3·47  |  10. ·238×·0347  |  16. ·00238×347000
5. 5·62×·213  |  11. ·0562×·0000213  |  17. ·00562×21300
6. ·278×·547  |  12. ·00278×·000547  |  18. 27800×·000547

**31.**

19. 98·7654 × ·983427
20. ·123456 × ·654321
21. 5·78934 × ·000763

22. ·007639 × 763900
23. 87·6591 × 684000
24. ·000009 × ·000983

25. 100 × ·01 × ·001 × ·0001 × 1000
26. 300 × ·003 × ·0003 × 3000 × ·00003
27. 5000 × 500 × ·0007 × ·035 × ·00005
28. ·003 × ·03 × ·3 × ·0003 × 30000

Find the following products by Vulgar and Decimal Fractions:

29. $\frac{3}{4} \times \frac{6}{25} \times 2\frac{1}{2}$
30. $\frac{7}{8} \times \frac{11}{20} \times \frac{3}{4}$
31. $5\frac{1}{3} \times \frac{18}{25} \times \frac{17}{51}$

32. $\frac{4}{5} \times \frac{9}{10} \times \frac{7}{8}$
33. $\frac{3}{8} \times \frac{7}{18} \times \frac{18}{40}$
34. $2\frac{3}{5} \times 1\frac{11}{25} \times \frac{3}{80}$

**32.** In the DIVISION of Decimals we divide as in integers, and point the quotient so that it may contain as many decimal places as are in the dividend, diminished by the number in the divisor.

(1) Divide 228·75 by 30·5; and 6·4 by 25·6.

```
30·5)228·75(7·5        25·6)6·400(·25
     2135                   512
     ────                   ────
     1525                   1280
     1525                   1280
```

$\frac{22875}{100} \times \frac{10}{305} = \frac{1}{10} \times \frac{22875}{305} = 7\cdot5.$  $\frac{64}{10} \times \frac{10}{256} = \frac{64}{256} = \cdot25.$

In dividing 228·75 by 30·5, since there are *two* places in the dividend and *one* in the divisor, we point off *one* in the quotient. In dividing 6·4 by 25·6, since we use *three* places in the dividend and *one* in the divisor, we point off *two* in the quotient.

The following examples illustrate various modifications of the general rule:—

(2) Divide 48·97 by ·59; and 292·3 by 3·95.

```
·59)48·97(83           3·95)292·30(74
    472                     2765
    ───                     ────
    177                     1580
    177                     1580
```

(3) Divide ·68625 by 91500; and 32·1 by 128400.

```
91500)·68625(·0000075    128400)32·100(·00025
      6405                      2568
      ────                      ────
      4575                      6420
      4575                      6420
```

Since in dividing ·68625 by 915 we would have ·00075, by increasing the divisor 100 times we diminish the quotient as many times, and thus obtain ·0000075. Similarly, in dividing 32·1 by 128400, the number of decimal places in the quotient is = the sum of the number of decimal places used in the dividend, and of the number of annexed ciphers in the divisor.

**32.**  (4) Divide 2230·1 by ·769; and 1400 by ·00224.

```
·769)2230·1(2900        ·00224)1400·00(625000
     1538                      1344
    ─────                     ─────
     6921                      560
     6921                      448
                             ─────
                              1120
                              1120
                             ─────
```

In dividing 2230·1 by 769 we would have the quotient 2·9. By diminishing the divisor 1000 times we increase the quotient as many times, and thus obtain 2900. Similarly, in dividing 1400 by ·00224, we annex as many ciphers to the quotient as there are decimal places in the divisor, diminished by the number of decimal places used in the dividend.

We may often find it of advantage to reduce the divisor to an integer, and move the decimal point in the dividend as many places towards the right as we do in the divisor.

According to this method, the examples in (2) and (4) would be expressed in the following manner:—

```
    59)4897(              769)2230100(
    395)29230(            224)140000000(
```

1. 1·7503 ÷ 7·61
2. 40·3858 ÷ 6·34
3. 39·538 ÷ ·53
4. 392·37 ÷ 31·9
5. 110·925 ÷ 1·53
6. 5·2441 ÷ 22·9

7. 1750·3 ÷ ·0761
8. 4038·58 ÷ ·0634
9. 3953·8 ÷ ·053
10. 39237 ÷ ·319
11. 1109·25 ÷ ·0153
12. 524·41 ÷ ·0229

13. 175·03 ÷ 76100
14. ·403858 ÷ 63400
15. ·39538 ÷ 5300
16. ·39237 ÷ 3190
17. ·110925 ÷ 153000
18. ·52441 ÷ 22900

19. ·0156366 ÷ ·0042
20. ·03486 ÷ 4·98
21. ·378816 ÷ 5·919

22. 20973·6 ÷ ·8739
23. 9110·64 ÷ 2900
24. 7·127577 ÷ 1·0053

Find the following quotients by Vulgar and Decimal Fractions:

25. $\frac{4}{5} \div 1\frac{1}{4}$
26. $7\frac{1}{2} \div \frac{3}{10}$

27. $1\frac{11}{14} \div 10\frac{2}{3}$
28. $4\frac{3}{8} \div 2\frac{2}{5}$

29. $\frac{11}{25} \div \frac{4}{15}$
30. $3\frac{1}{8} \div 12\frac{2}{5}$

**33.**  In an INTERMINATE DECIMAL, one figure or a series of figures continuously recurs. The figures which recur form a *Period*. When the decimal contains the recurring period only, it is termed a *Pure Interminate*, as ·333, &c., written ·3̇; ·036036, &c., written ·0̇3̇6̇. When the decimal contains a terminate as well as an interminate part, it is termed a *Mixed Interminate*, as ·1666, &c., written ·16̇; ·159090, &c., written ·159̇0̇. When the period contains one figure, the decimal is called a *Repeater*; but when more than one, it is called a *Circulator*.

**33.**

| PURE INTERMINATE. | MIXED INTERMINATE. |
|---|---|
| Pure Repeater.........as...·$\dot{3}$ | Mixed Repeater.....as...·1$\dot{6}$ |
| Pure Circulator.......″ ...·$\dot{0}3\dot{6}$ | Mixed Circulator.... ″ ...·15$\dot{9}\dot{0}$ |

A vulgar fraction whose denominator in the lowest terms of the fraction contains neither of the prime factors 2 or 5, produces a pure interminate; thus, $\frac{1}{3} = \cdot\dot{3}$; $\frac{3}{7} = \cdot\dot{4}2857\dot{1}$.

A vulgar fraction whose denominator in the lowest terms of the fraction contains 2 or 5, and one or more of the other primes, produces a mixed interminate; thus, $\frac{1}{6} = \cdot1\dot{6}$; $\frac{144}{275} = \cdot52\dot{3}\dot{6}$.

Express the following vulgar fractions as decimals:

    (1) $\frac{6}{7} = \cdot\dot{8}5714\dot{2}$.

By annexing ciphers to 6 and dividing by 7, we find that the quotient consists of a period of six figures.

    (2) $\frac{7}{22} = \cdot3\dot{1}\dot{8}$.

The interminate part of the decimal begins at the second place, and consists of a period of two figures.

    (3) $\frac{1}{17} = \cdot\dot{0}588235294117647$.

When the numerator is unity, and the denominator such a prime as will produce a considerable number of figures in the period, we may work as follows: By taking out the decimal, say to 5 places, we obtain $\frac{1}{17} = \cdot05882\frac{6}{17}$, which, multiplied by 6, gives the decimal for $\frac{6}{17}$. Proceeding similarly with the other final vulgar fractions, as in the subjoined process, we have $\frac{1}{17} = \cdot05882352941176470588\frac{6}{17}$. By examining where the figures begin to recur, we obtain a period of sixteen figures as above.

$\frac{1}{17} = \cdot05882\frac{6}{17}$
$\frac{6}{17} = \cdot35294\frac{2}{17}$
$\frac{2}{17} = \cdot11764\frac{12}{17}$
$\frac{12}{17} = \cdot70588\frac{4}{17}$

| | | | |
|---|---|---|---|
| 1. $\frac{2}{3}$ | 6. $\frac{1}{19}$ | 11. $\frac{5}{6}$ | 16. $\frac{64}{225}$ |
| 2. $\frac{7}{9}$ | 7. $\frac{1}{37}$ | 12. $\frac{7}{26}$ | 17. $\frac{15}{296}$ |
| 3. $\frac{4}{11}$ | 8. $\frac{4}{27}$ | 13. $\frac{12}{22}$ | 18. $\frac{13}{56}$ |
| 4. $\frac{5}{13}$ | 9. $\frac{7}{143}$ | 14. $\frac{3}{27}$ | 19. $\frac{11}{74}$ |
| 5. $\frac{11}{21}$ | 10. $\frac{3}{271}$ | 15. $\frac{7}{24}$ | 20. $\frac{5}{108}$ |

**34.** Express the following interminate decimals as vulgar fractions:

    (1) ·$\dot{1}8\dot{5}$.

$$1000 \times \cdot\dot{1}8\dot{5} = 185\cdot\dot{1}8\dot{5}$$
$$1 \times \cdot\dot{1}8\dot{5} = \phantom{185}\cdot\dot{1}8\dot{5}$$

Therefore, $999 \times \cdot\dot{1}8\dot{5} = 185$

And, $\cdot\dot{1}8\dot{5} = \frac{185}{999} = \frac{5}{27}$

In reducing a pure interminate to the form of a vulgar fraction, we take the period as the numerator, and write "9" as often in the denominator as there are figures in the period.

**34.**

(2) $\cdot\dot{4}8\dot{1} = \frac{481}{999} = \frac{13}{27}$.

(3) $\cdot0\dot{7}692\dot{3} = \frac{76923}{999999} = \frac{1}{13}$.

| | | | |
|---|---|---|---|
| 1. $\cdot\dot{4}$ | 5. $\cdot1\dot{3}\dot{5}$ | 9. $\cdot2\dot{9}\dot{6}$ | 13. $\cdot\dot{4}2857\dot{1}$ |
| 2. $\cdot\dot{5}\dot{4}$ | 6. $\cdot2\dot{8}\dot{8}$ | 10. $\cdot0\dot{2}3\dot{1}$ | 14. $\cdot\dot{1}5384\dot{6}$ |
| 3. $\cdot\dot{0}\dot{7}$ | 7. $\cdot2\dot{5}\dot{9}$ | 11. $\cdot003\dot{6}\dot{9}$ | 15. $\cdot\dot{0}0040\dot{7}$ |
| 4. $\cdot\dot{9}6\dot{2}$ | 8. $\cdot\dot{4}8\dot{1}$ | 12. $\cdot0243\dot{9}$ | 16. $\cdot\dot{0}4761\dot{9}$ |

(4) $\cdot6\dot{8}\dot{1}$.

$$1000 \times \cdot6\dot{8}\dot{1} = 681\cdot\dot{8}\dot{1}$$
$$10 \times \cdot6\dot{8}\dot{1} = 6\cdot\dot{8}\dot{1}$$

Therefore, $990 \times \cdot6\dot{8}\dot{1} = 675$

And, $\cdot6\dot{8}\dot{1} = \frac{675}{990} = \frac{15}{22}$.

In reducing a mixed interminate to the form of a vulgar fraction, we take for the numerator the difference between the integral numbers, which respectively contain the figures of the decimal and those of its terminate part; and for the denominator we write "9" as often as there are figures in the period, and annex as many ciphers as there are figures in the terminate part.

(5) $\cdot12\dot{3}\dot{4} = \frac{1234-12}{9900} = \frac{1222}{9900} = \frac{611}{4950}$.

The following method may also be employed in reducing a mixed interminate to the form of a vulgar fraction :—

$$\cdot6\dot{8}\dot{1} = \cdot6 + \cdot0\dot{8}\dot{1} = \tfrac{6}{10} + \tfrac{1}{10} \text{ of } \cdot\dot{8}\dot{1}$$
$$= \tfrac{6}{10} + \tfrac{1}{10} \text{ of } \tfrac{81}{99} = \tfrac{6}{10} + \tfrac{81}{990} = \frac{594+81}{990}$$
$$= \tfrac{675}{990} = \tfrac{15}{22}.$$

| | | | |
|---|---|---|---|
| 17. $\cdot1\dot{6}$ | 21. $\cdot70\dot{4}\dot{5}$ | 25. $\cdot005\dot{4}$ | 29. $\cdot009\dot{6}\dot{2}$ |
| 18. $\cdot11\dot{6}$ | 22. $\cdot00\dot{4}\dot{5}$ | 26. $\cdot091\dot{6}$ | 30. $\cdot00021\dot{6}$ |
| 19. $\cdot013\dot{8}$ | 23. $\cdot00\dot{5}\dot{4}$ | 27. $\cdot091\dot{6}$ | 31. $\cdot\dot{5}14285\dot{7}$ |
| 20. $\cdot41\dot{6}$ | 24. $\cdot005\dot{4}$ | 28. $\cdot091\dot{6}$ | 32. $\cdot1\dot{0}7692\dot{3}$ |

**35.**

(1.) $\cdot\dot{3} + \cdot\dot{8}\dot{1} + \cdot\dot{0}3\dot{7} + \cdot375$.

Since the terminate decimal ·375 occupies three places, the interminate part of the sum begins at the fourth place. The periods, consisting of 1, 2, and 3 figures respectively, are extended 6 places beyond the terminate decimal, and as they then recur in the same relative order, the period in the sum thus consists of 6 places, which is the L. C. M. of 1, 2, 3.

$$\begin{array}{l} \cdot333|\dot{3}3333\dot{3} \\ \cdot818|\dot{1}8181\dot{8} \\ \cdot037|\dot{0}3703\dot{7} \\ \cdot375| \\ \hline 1\cdot563\,\dot{5}5218\dot{8} \end{array}$$

In extending periods to as many places as are denoted by the L. C. M. of the number of places in each, we are said to make the periods *similar*.

**35.** In the Addition of Interminate Decimals, having extended the Interminates to the longest terminate part, we make the periods similar and then find the sum.

(2) $\cdot\dot{3} + \cdot\dot{4} + \cdot\dot{7} = 1\cdot\dot{5}.$

$3 + 4 + 7 = 14.$ Since, by extending the decimals a place to the right, we would obtain the same sum, we add in 1, and thus obtain the sum $= 1\cdot\dot{5}.$

(3) $4\cdot9\dot{6}\dot{2} + \cdot41\dot{6} + 5\cdot07692\dot{3}.$

As the periods have been made similar, we first add the columns at the beginning of the similar periods to find the number to be carried to the last column.

```
 4·96 296296
 ·41 666666
 5·07 692307
─────────────
10·45 655270
```

(4) $\cdot\dot{3} + \cdot\dot{4} + \cdot\dot{5} + \cdot\dot{6} = 2.$

$3 + 4 + 5 + 6 = 18,$ so with the carrying figure the sum is $= 1\cdot\dot{9} = 1 + \frac{9}{9} = 2.$

When we obtain 9 as a repeater, we write 0 and carry 1.

1. $\cdot\dot{5} + \cdot\dot{1} + \cdot\dot{6} + \cdot\dot{3}$
2. $\cdot\dot{2} + \cdot\dot{8} + \cdot\dot{7} + \cdot\dot{4} + \cdot\dot{6}$
3. $\cdot0\dot{9} + \cdot4\dot{5} + \cdot2\dot{7} + \cdot5\dot{4}$
4. $\cdot3\dot{6} + \cdot1\dot{8} + \cdot6\dot{3} + \cdot8\dot{1}$
5. $\cdot9\dot{6}\dot{2} + \cdot2\dot{6}\dot{1} + \cdot1\dot{6}\dot{2} + \cdot1\dot{8}\dot{5}$
6. $\cdot3\dot{7}\dot{0} + \cdot2\dot{5}\dot{9} + \cdot0\dot{3}\dot{6} + \cdot4\dot{0}\dot{7}$

7. $\cdot5\dot{0}\dot{9} + \cdot0\dot{3}\dot{7} + \cdot75$
8. $\cdot21\dot{6} + \cdot21\dot{6} + \cdot2\dot{1}\dot{6} + \cdot21\dot{6}$
9. $\cdot0\dot{3}\dot{7} + \cdot5\dot{0}\dot{3} + \cdot142857$
10. $\frac{2}{27} + \frac{3}{37} + \frac{1}{111} + \frac{8}{333}$
11. $\frac{3}{7} + \frac{11}{36} + \frac{3}{61} + \frac{1}{420}$
12. $\frac{1}{117} + \frac{8}{3367} + \frac{4}{5191} + \frac{10}{207}$

**36.** In Subtracting an Interminate Decimal from another, we make the periods similar, and then find the difference.

(1) $\cdot\dot{9}14285\dot{7} - \cdot3\dot{9}\dot{6}\dot{2}.$

Having found that we carry 1 from the beginning of the period, we take 3 from 7, &c.

```
·9 142857
·3 962962
─────────
·5 179894
```

(2) $\cdot275 - 1\dot{9}6\dot{2}.$

In subtracting an Interminate from a Terminate, instead of carrying from the beginning of the period, we may subtract each of the figures in the interminate from 9; thus, having obtained 703 by taking 296 from 999, we carry 1 to 6 in the subtrahend.

```
·275
·196 296
─────────
·078 703
```

1. $\cdot1\dot{6} - \cdot0\dot{7}$
2. $\cdot21\dot{6} - \cdot158\dot{3}$
3. $\cdot24\dot{3} - \cdot07\dot{4}$
4. $\cdot07692\dot{3} - \cdot0375$

5. $\cdot2\dot{3}\dot{4} - \cdot167\dot{2}$
6. $\cdot28571\dot{4} - \cdot00\dot{9}\dot{3}$
7. $\cdot306 - \cdot00\dot{9}$
8. $\cdot00\dot{3} - \cdot000\dot{3}$

9. $\cdot0\dot{3}\dot{0} - \cdot030\dot{0}$
10. $\frac{1}{101} - \frac{1}{233}$
11. $\frac{1}{55} - \frac{1}{813}$
12. $\frac{1}{37} - \frac{1}{57067}$

**37.** In Multiplying an Interminate Decimal by a Terminate, we proceed in the following manner:—

(1) ·$\dot{7}$62$\dot{3}$ × 27·5.

In multiplying by 5, we carry 3 from the beginning of the period; similarly, in multiplying by 7, we carry 5; and in multiplying by 2, we carry 1. We then extend the periods, and find the sum.

$$\begin{array}{r} ·\dot{7}62\dot{3} \\ 27·5 \\ \hline 38118 \\ 533663 \\ 1524752 \\ \hline 20·9\dot{6}534 \end{array}$$

| | | |
|---|---|---|
| 1. ·108$\dot{3}$ × 4 | 5. ·$\dot{9}$6$\dot{2}$ × 11 | 9. 3·0$\dot{9}$ × 37 |
| 2. ·21$\dot{6}$ × 7 | 6. ·7$\dot{5}\dot{3}$ × 64 | 10. ·0$\dot{3}\dot{7}$ × 23 |
| 3. ·3$\dot{2}$ × 9 | 7. 8·4$\dot{6}$ × 846 | 11. ·003$\dot{3}$ × 606 |
| 4. ·$\dot{1}$4285$\dot{7}$ × 5 | 8. 7·$\dot{2}\dot{7}$ × 72 | 12. ·0975$\dot{6}$ × 250 |

In Dividing an Interminate Decimal by a Terminate, we extend the dividend until the quotient recurs.

(2) ·$\dot{1}$4$\dot{8}$ ÷ 12.

$$12)·148148148$$
$$\overline{·\dot{0}1234567\dot{9}}$$

| | | |
|---|---|---|
| 13. ·$\dot{8}$5714$\dot{2}$ ÷ 6 | 15. ·$\dot{0}$23$\dot{1}$ ÷ 308 | 17. 24·10$\dot{6}$ ÷ 32 |
| 14. ·$\dot{0}$352$\dot{3}$ ÷ 26 | 16. 57·1$\dot{8}$ ÷ 37 | 18. 33·$\dot{3}$ ÷ 271 |

In Multiplying or Dividing by an Interminate Decimal, we reduce it to a vulgar fraction.

(3) ·$\dot{0}$7692$\dot{3}$ × ·$\dot{2}$85714.

$$·\dot{2}85714 = \tfrac{2}{7}$$

$$\begin{array}{r} ·\dot{0}7692\dot{3} \\ 2 \\ \hline 7)·153846 \\ \hline ·\dot{0}2197\dot{8} \end{array}$$

(4) ·5$\dot{3}\dot{6}$ ÷ ·$\dot{5}$.

$$·\dot{5} = \tfrac{5}{9}$$

$$\begin{array}{r} ·5\dot{3}\dot{6} \\ 9 \\ \hline 5)4·8272 \\ \hline ·96\dot{5}\dot{4} \end{array}$$

We may sometimes reduce both the multiplier and the multiplicand, or both the dividend and the divisor, to vulgar fractions.

| | |
|---|---|
| 19. ·$\dot{2}\dot{7}$ × ·$\dot{3}$ | 25. ·$\dot{0}$036$\dot{9}$ ÷ ·$\dot{0}$027$\dot{1}$ |
| 20. ·$\dot{0}$3$\dot{7}$ × ·$\dot{0}$2$\dot{7}$ | 26. ·$\dot{0}$243$\dot{9}$ ÷ ·$\dot{1}$707$\dot{3}$ |
| 21. ·$\dot{0}$243$\dot{9}$ × ·$\dot{0}$036$\dot{9}$ | 27. $\frac{1}{360}$ × ·0$\dot{3}$ |
| 22. 2·2503$\dot{7}$ × ·4$\dot{1}\dot{8}$ | 28. $\frac{1}{27}$ × $\frac{1}{37}$ |
| 23. 10 ÷ ·$\dot{3}$ | 29. ·$\dot{0}$081$\dot{3}$ ÷ $\frac{1}{41}$ |
| 24. 23 ÷ 2·0$\dot{9}$ | 30. $\frac{6}{13}$ ÷ 1$\frac{2}{7}$ |

**38.** If we wish to have 3·141159265358979, &c. correct to 8 decimal places, we take 3·14159265; but if we desire to carry it

**38.** to 4 places merely, it will be more accurate to write 3·1416 than 3·1415, for the fifth decimal place being above 5, the former is nearer to the true decimal than the latter, and is thus a nearer APPROXIMATION.

(1) Give approximations to ·8450980400+ from 9 places to 1 place successively.

·845098040+; ·84509804+; ·8450980+; ·845098+
·84510—; ·8451—; ·845+; ·85—; ·8+.

By affixing "+" we mean that the true value of the decimal is > the approximation; and by affixing "—", that the former is < the latter.

If ·8450980400 had been terminate, we would have written ·84509804 merely. But if we had written only eight places in the approximate decimal, it would seem as if we knew not the next two.

Give approximations to the following from 9 places to 1 place successively.

1. ·0413926852—  |  3. ·4771212547+  |  5. ·6989700043+
2. ·3010299957—  |  4. ·6020599913+  |  6. ·7781512504—

(2) Find the sum ·428571̇ + ·39024̇, to 6 decimal places.

We extend the decimals to 7 places, and finding the sum of the 7th column, we add in the carriage to the 6th, and thus obtain the sum correct to 6 places.

·4285714+
·3902439+
―――――
·818815+

(3) Find the sum 1·05̇ + ·571428̇ + ·83̇ + ·39024̇, to 4 decimal places.

To obtain the last figure as the nearest approximation, it is often necessary to extend the decimals two places beyond the number required. The sum of the 5th column, increased by the carriage from the 6th, being nearer 20 than 10, we carry 2 to the 4th column.

1·055556—
·571429—
·833333+
·390244—
―――――
2·8506—

7. 7·2̇7̇ + 9·291̇6̇ + 8·3̇6̇ to 3 pl.

8. ·0̇3̇6̇ + ·0̇3̇6̇ + ·0̇3̇6̇ to 4 pl.

9. ·0̇243̇9̇ + ·0̇0̇3̇ + 3·1416– to 4 pl.

10. ·91908– + ·72428– + ·72607+ to 5 pl.

11. $\frac{2}{17} + \frac{3}{19} + \frac{4}{23}$ to 6 pl.  |  13. $3\frac{1}{5} + 7\frac{3}{13} + 2\frac{2}{37}$ to 3 pl.

12. $\frac{2}{13} + \frac{3}{7} + \frac{6}{11}$ to 4 pl.  |  14. $\frac{2}{3}$ of $\frac{10}{17} + \frac{3}{4}$ of $\frac{10}{19} + 7\frac{1}{33}$ to 3 pl.

(4) From ·12195̇ subtract ·0693̇ to 5 places.

·121951+
·069307—
―――――
·05264+

(5) From ·142857̇ subtract ·00813̇ to 5 places.

**38.** The extra figure in the 6th place of the remainder being $>$ 5, we increase the figure in the 5th place by 1.

$$\begin{array}{r} \cdot142857+ \\ \cdot008130+ \\ \hline \cdot13473- \end{array}$$

(6) From $\cdot01298\dot{7}$ subtract $\cdot0027\dot{1}$ to 8 places.

Since the extra figure is $>$ 5, we cancel the carriage.

$$\begin{array}{r} \cdot012987013- \\ \cdot002710027+ \\ \hline \cdot01027699- \end{array}$$

15. $\cdot\dot{7} - \cdot7291\dot{6}$ to 4 pl.

16. $\cdot\dot{2}5\dot{9} - \cdot\dot{0}027\dot{1}$ to 5 pl.

17. $\cdot\dot{9}6\dot{2} - \cdot\dot{9}\dot{0}$ to 4 pl.

18. $\cdot0625 - \cdot041\dot{6}$ to 3 pl.

19. $\cdot2 - \cdot008\dot{3}$ to 3 pl.

20. $1\cdot041393 - - \cdot698970+$ to 5 pl.

21. $1\cdot41497+ - 1\cdot32222-$ to 4 pl.

22. $\frac{1}{355} - \frac{1}{371}$ to 5 pl.

23. $\frac{1}{77} - \frac{1}{333}$ to 5 pl.

24. $2\frac{1}{4} - \frac{6}{13} + \frac{2}{513} - \frac{1}{407}$ to 5 pl.

**39.** In CONTRACTED MULTIPLICATION we obtain a product which is correct to a certain number of places.

If we wish to find the product of the terminate decimals 5·2467 and 4·2635 to *four* decimal places merely, it is evident that the figures to the right of the line in A are unnecessary.

In B, we commence the first line by multiplying the figure in the fourth place having the local value of 7 *ten thousandths* by 4

| | A | B |
|---|---|---|
| | 5·2467 | 5·2467 |
| | 4·2635 | 4·2635 |
| | 26\|2335 | 209868 |
| | 157\|401 | 10493 |
| | 3148\|02 | 3148 |
| | 10493\|4 | 157 |
| | 209868\| | 26 |
| | 22·3693\|0545 | 22·3692 |

*units;* the second line by multiplying 6 *thousandths* in the third place by 2 *tenths*, adding in the carriage of 1 from $2 \times 7$; the third line by multiplying 4 *hundredths* in the second place by 6 hundredths, adding in the carriage of 4 from $6 \times 6$, &c. Since the first column on the right has the local value of *ten thousandths*, there are thus four decimal places in the product, as required.

To insure accuracy in the last decimal place of the approximate product, we work for *one* place *more* than what is required. To accommodate the eye, we *invert the multiplier,* and put its units' place under the place in the multiplicand whose local value is the same as that of the last decimal place for which we are working.

(1) Multiply 5·2467 by 4·2635 to 4 places.

Working for *five* places, we invert the multiplier, and put the figure in the *units'* place of the multiplier under the *fifth* place of the multiplicand.

In adding, we carry 1 from the last column.

$$\begin{array}{r} 5\cdot24670 \\ Inv.\ (5362\cdot4) \\ \hline 2098680 \\ 104934 \\ 31480 \\ 1574 \\ 262 \\ \hline 22\cdot3693 \end{array}$$

**39.**    (2)  Multiply  ·0243$\dot{9}$  by  ·03$\dot{7}$  to 8 places.

Working for *nine* places, we place the inverted multiplier so that its units' place may be under the *ninth* of the multiplicand. We carry 1 from the last column, and as there are five significant figures in the product, we prefix three ciphers.

$$\begin{array}{r} ·02439024 \\ Inv.\ (40730730·) \\ \hline 731707 \\ 170732 \\ 732 \\ 170 \\ 1 \\ \hline ·00090334 \end{array}$$

1. 4·5625 × 3·375 to 5 pl.

2. 5·7563 × 3·996 to 3 pl.

3. 69·235 × 2·525 to 3 pl.

4. 14·36738 × 30·61725 to 5 pl.

5. ·0842367×52·6739 to 6 pl.

6. ·74216 × ·8237 to 5 pl.

7. 4·0243$\dot{9}$ × ·50$\dot{2}\dot{7}$ to 5 pl.

8. 5·85714$\dot{2}$ × 8·0$\dot{9}$ to 5 pl.

Let us find the product of the approximate factors 324·1674± and 2·12967±. The former may stand for any number between 324·16735 and 324·16745; and the latter for any between 2·129665 and 2·129675. Since the product of the least values 324·16735 and 2·129665 = 690·367859, and that of the greatest values 324·16745 and 2·129675 = 690·371314, the product of the approximate factors can therefore be guaranteed to two decimal places only, as 690·37±.

As the factors in the accompanying process are approximate, we see that of the *nine* decimal places in the product *seven* are indeterminate. The number of determinate places is=9—7; 9 = 5+4, the sum of the numbers of decimal places in the factors; 7, corresponding

$$\begin{array}{r} 324·1674\ * \\ 2·12967\ * \end{array}$$

| | | |
|---:|:---|:---|
| | * * * * * * * | * * |
| 2 | 2 6 9 1 7 1 8 | * |
| 1 9 | 4 5 0 0 4 4 * | |
| 2 9 1 | 7 5 0 6 6 * | |
| 3 8 9 0 | 0 0 8 8 * | |
| 6·4 8·8 3 | 4 8 * | |
| 6 9 0·3 7 | * * * * * * * | |

to the number of figures in the factor 324·1674±, is = 3+4, the sum of the numbers of integral and decimal places in that factor. By cancelling the number of decimal places in the factor having the greater number of figures, we have 9 — 7 = 5 — 3. The number of determinate places is = the number of *decimal* places in the factor having the *fewer* figures diminished by the number of *integral* places in the other.

(3)  Find the product of 31·7436± by ·76321± to as many places as can be depended on.

Since the number of decimals in the factor having the fewer figures is = 5, and the number of integral places in the other is = 2, the number of determinate places is = 5 — 2 = 3. We therefore work for 4 places.

$$\begin{array}{r} 31·7436 \\ Inv.\ (12367·) \\ \hline 222205 \\ 19046 \\ 952 \\ 63 \\ 3 \\ \hline 24·227 \end{array}$$

**39.** Find the following products to as many places as can be depended on :—

9. $2 \cdot 183 \pm \times \cdot 00704 \pm$.

☞ ·00704, whose significant figures extend over *three* places, has in all *five* decimal places. The other factor contains *one* integral place. The number of determinate places $= 5 - 1$.

10. $\cdot 000732 \pm \times 2 \cdot 8 \pm$.

☞ In ·732 $\pm \times 2 \cdot 8 +$, the number of reliable places would $= 1$. Since we have ·000732 $\pm$ as a factor, we remove the point *three* places to the left, and thus increase the number of reliable places.

11. $\cdot 23 \pm \times 7142 \cdot 3 \pm$.

☞ The number of decimal places in the factor having the fewer figures being less by 2 than the number of integral places in the other, we cannot depend on the last 2 integral places of the product, and thus can give it in *hundreds* only.

12. $1 \cdot 375 \times \cdot 2304 \pm$.

When one of the factors is terminate, the number of determinate places is $=$ the number of decimal places in the approximate factor, diminished by the number of integral places in the terminate.

13. $17 \cdot 69235 \pm \times 2 \cdot 00976 \pm$    17. $\cdot 7854- \times \cdot 0036712 \pm$
14. $16 \cdot 3467 \pm \times 8 \cdot 3146 \pm$    18. $\cdot 052 \pm \times 12345 \cdot$
15. $3 \cdot 247 \pm \times \cdot 00603 \pm$    19. $\cdot 275 \times 3 \cdot 2463 \pm$
16. $3 \cdot 1416- \times \cdot 007009 \pm$    20. $2 \cdot 005 \times \cdot 00017 \pm$

**40.** In CONTRACTED DIVISION, we obtain a quotient which is correct to a certain number of places.

(1) Let us divide $74 \cdot 0625$ by $\cdot 3147$, of which both are terminate, so as to obtain *three* decimal places in the quotient.

By inspection, we find by dividing 74 by ·3, that there will be *three* integral places in the quotient. We thus require *six* figures in the quotient. Annexing as many ciphers to the divisor as make it contain *six* figures, we find the first figure in the quotient, and then elide a figure from the divisor at each successive step. We need not write the ciphers in the first two partial products.

```
·3,1,4,7,0,0) 74·0625 ( 235·343 +
              6294
              11122
              9441
              16815
              15735
              1080
               944
               136
               126
                10
                 9
```

**40.**    (2) Divide ·0125 by 30·725, to obtain *seven* decimal places in the quotient.

We find that as there will be *three* prefixed ciphers in the dividend, the number of figures necessary to make up the required number in the quotient will be 7 — 3, or *four*. We commence to divide by 3072, and elide a figure at each successive step.

$$30·72{,}5 ) ·012500 ( ·0004068+$$
$$\underline{12290}$$
$$\underline{210}$$
$$184$$
$$\overline{26}$$
$$25$$
$$\overline{\phantom{25}}$$

To obtain a certain number of figures in the quotient of two terminate decimals, we begin the division by having as many figures in the divisor as are = the number of required decimal places increased by the number of integral places in the quotient, or diminished by the number of prefixed ciphers in it. We then continue to elide a figure from the divisor at each successive step until it is exhausted.

Find the quotients of the following numbers having terminate decimals :—

1. 6·75 ÷ 3·25 to 4 pl.
2. 10· ÷ 4·75 to 3 pl.
3. 20·6 ÷ 3·3125 to 3 pl.
4. 6·23475 ÷ ·04875 to 3 pl.
5. 4·12189 ÷ ·04763 to 2 pl.
6. ·004365 ÷ ·71215 to 5 pl.
7. ·0007 ÷ 3·125 to 6 pl.
8. ·00034625 ÷ 631·247 to 10 pl.

When either the divisor or the dividend, or both, are approximate, we can depend on only a certain number of places in the quotient, as may be seen in the following examples :—

(3)    2·5 ÷ ·0773±.
·0773) 2·500 (32·3+
    2319
    ─────
    181
    155
    ───
    26
    23
    ──

(5)    6·143+ ÷ ·007354±.
·00735|4) 6·143 (835·+
    5883
    ────
    260
    221
    ───
    39
    37
    ──

(4)    ·0031416- ÷ ·67±.
·67) ·00314|16( ·0047—
    268
    ───
    46
    47
    ──

(6)    ·007316± ÷ 7·4.
7·4) ·007316 ( ·000989 —
    666
    ───
    656
    592
    ───
    64
    67
    ──

**40.** In all cases, we first find the initial figure in the quotient and point it.

When the divisor is approximate, and the dividend has more determinate places than are in the divisor, as in (3) and (4), we begin to elide the figures in the divisor after the first partial product. When, as in (5), the dividend is approximate, and the divisor can produce more determinate places than are in the dividend, as many figures only of the divisor must be taken as will make the first partial product contain no more than are in the dividend. But when, as in (6), the divisor is terminate, and has its significant figures extending over fewer places than the number of the determinate in the dividend, we carry on the division in the ordinary way till the dividend is exhausted, and then commence the contraction.

9. $\cdot7234\pm \div \cdot525$  
10. $\cdot00313\pm \div 7\cdot4$  
11. $1\cdot0367\pm \div \cdot94364\pm$  
12. $12\cdot3\pm \div \cdot8738\pm$  
13. $2\cdot575 \div \cdot234\pm$  
14. $10\cdot \div \cdot5236-$  
15. $5\cdot2673\pm \div \cdot06731\pm$  
16. $2\cdot0167\pm \div \cdot733\pm$  
17. $1\cdot0035 \div \cdot0417\pm$  
18. $10\cdot \div 21\cdot63\pm$  
19. $1\cdot \div 2\cdot302585093-$  

20. $4\cdot \div 2\cdot167\pm$  
21. $\cdot1 \div \cdot000767\pm$  
22. $72\cdot1 \div \cdot00312\pm$  
23. $10\cdot \div \cdot000763\pm$  
24. $\cdot007635 \div 7\cdot142\pm$  
25. $\cdot073167\pm \div 2\cdot25$  
26. $1\cdot \div 12\cdot56637+$  
27. $42\cdot75 \div \cdot00077\pm$  
28. $630\cdot \div \cdot0739\pm$  
29. $\cdot0125 \div 71\cdot23\pm$  
30. $10\cdot \div 2\cdot718281828+$  

**41.** In REDUCING a simple quantity to the decimal of another in a higher name, we annex ciphers to the number of units in the former, and divide by the number which shows how often a unit of the lower name is contained in one of the higher.

(1) Reduce 9d. to the decimal of 1l.

We thus change $\frac{9}{12}$ to a decimal.

$$\begin{array}{r} \text{d.} \\ 12\,)\,9\cdot00 \\ \hline \cdot75\text{s.} \end{array}$$

In reducing a compound quantity to the decimal of a simple quantity, we reduce the number in the lowest name to the decimal of the next higher, to which we prefix the integer in the latter, and so proceed till we obtain the decimal of the required name.

(2) Reduce 4 lb. 7 oz. 15 dwt. to the decimal of 40 lb.

The accompanying process is equivalent to the following:—

15 dwt. $= \frac{15}{20}$ oz. $= \cdot75$ oz.

7·75 oz. $= \frac{7\cdot75}{12}$ lb. $= \cdot64583$ lb.

4·64583 lb. $= \frac{4\cdot64583}{40} = \cdot11614583$ of 40 lb.

$$\begin{array}{r|l} & \text{dwt.} \\ 20 & 15\cdot \\ 12 & 7\cdot75 \text{ oz.} \\ 40 & 4\cdot64583 \text{ lb.} \\ \hline & \cdot11614583 \end{array}$$

**41.** When the quantities are expressed in mixed numbers containing vulgar fractions or decimals, we proceed as follows :

(3) Reduce 4¾ min. to the decimal of 15·2 hours.

$$60 ) \underline{4·75 \text{ min.}}$$
$$15·2 ) \overline{·07916} \text{ ho. (·00520833̇.}$$

We may sometimes *cancel* thus :—

$$\frac{4·75}{60 \times 15·2} = \frac{·25}{60 \times ·8} = \frac{·25}{48} = ·0052083̇.$$

In reducing a compound quantity to the decimal of another, we find the vulgar fraction, which shows what part the former is of the latter, and reduce it to a decimal.

(4) Reduce £2 ⫫ 11 ⫫ 8 to the decimal of £5 ⫫ 7 ⫫ 7½.

By the method of § 20., No. (4); £2⫫11⫫8 = $\frac{1210}{2535}$ of £5⫫7⫫7½

$$\frac{1210}{2535} = ·48006194+$$

*Otherwise :* By reducing 11/8 to the decimal of £1, and prefixing 2, we obtain £2·583̇, and are thus said to have reduced £2⫫11⫫8 to the decimal of £1. Similarly, £5⫫7⫫7½ reduced to the decimal of £1 = £5·38125.

$$£2·583̇ \div £5·38125 = ·48006194+$$

1. 8d..................1s.
2. 15 cwt..............1 T.
3. 30 in...............1 yd.
4. 7/6..................£1.
5. 13/4½...............£1.
6. 5/6¼.................£1.
7. 8 oz. 3 dwt.......1 lb. troy.
8. 3 fu. 10 po........1 ml.
9. 2 ro. 30 po........1 ac.
10. 3 qr. 15¾ lb.....10 cwt.
11. 3 bu. 3½ pk.......5 qr.
12. 6 ho. 9¼ min......3 da.
13. 2/8¼..............5/3¾
14. 7/8½..............15/3
15. 6/7¼..............18/9
16. 3 oz. 5 dwt.....1 lb. 3 oz.
17. 2 bu. 3 pk.....5 bu. 1 pk.
18. 2 ft. 3 in.......3 yd. 2 ft.
19. 5 fu. 8 po......7 fu. 20½ po.
20. 5min. 16½ sec..3 ho. 15 min.
21. 23° 27′ 37″.....90°.
22. 5 cwt. 3 qr.....2 T. 10 cwt.
23. 3 da. 10½ ho....3 wk. 4 da.
24. 6¾ min..........7 ho. 30 min.

25. From Delhi to Bombay the direct distance is 720 miles; and from Delhi to Madras, 1080 miles. Reduce the former to the decimal of the latter.

26. Westminster Hall is 270 feet long and 75 feet broad. Reduce the latter to the decimal of the former.

27. Reduce a sidereal day, which is = 23 ho. 56 min. 4·09 sec., to the decimal of a solar day of 24 hours.

28. Reduce the sidereal day of Jupiter, which is = 9 ho. 55 min.

**41.** 50 sec., to the decimal of the Earth's sidereal day, which is 23 ho. 56 min. 4·09 sec.

29. Reduce a solar year, which is = 365 da. 5 ho. 48 min. 49·7 sec., to the decimal of a sidereal year, which is = 365 da. 6 ho. 9 min. 9·6 sec.

30. Express the height of the Peak of Teneriffe, which is = 12232 feet, as the decimal of a mile.

31. Express £3 ″ 17 ″ 10½, the value of 1 lb. troy of sterling gold, in the decimal of £1.

32. The Danube is 1630 miles long, and from the source of the Missouri to the mouth of the Mississippi the distance is 4000 miles. Reduce the former to the decimal of the latter.

33. Reduce the weight of a Cologne mark, which is = 3608 grains, to the decimal of 1 lb. troy and of 1 lb. avoir.

**42.** In finding the value of a decimal of a unit, we multiply the decimal by the number of times the given unit contains the next lower unit, and so on as far as may be required.

(1) Find the value of £·7895.

$$£·7895$$
$$20$$
$$\overline{\phantom{0}}$$
$$s.\ 15·7900$$
$$12$$

£·7895 =        d. $\overline{9·48}$
15/9½ ²³₁₈        4
                 f. $\overline{1·92}$

(2) Find the value of ·583 oz. troy.

$$oz.\ ·58\dot{3}$$
$$20$$
$$\overline{\phantom{0}}$$
$$dwt.\ 11·\dot{6}$$
$$24$$
$$\overline{\phantom{0}}$$
$$gr.\ 16·$$

By multplying the interminate decimals, we obtain

·58$\dot{3}$ oz. = 11 dwt. 16 gr.

The following examples afford additional illustration of finding the values of decimals :—

(3) Find the value of 2·75 of 5·45 acres.

$$5·45$$
$$2·75$$
$$\overline{\phantom{0}}$$
$$2725$$
$$3815$$
$$1090$$
$$\overline{\phantom{0}}$$
$$ac.\ 14·9875$$
$$4$$
$$\overline{\phantom{0}}$$
$$ro.\ 3·9500$$
$$40$$
$$\overline{\phantom{0}}$$
$$po.\ 38·00$$

(4) Find the value of 2·425 of 5 cwt. 3 qr. 16 lb.

5 cwt. 3 qr. 16 lb. = 660 lb.

$$2·425$$
$$660$$
$$\overline{\phantom{0}}$$
$$14550$$
$$14550$$

28 | 1600·5
4 | ‾‾57 qr. 4 lb.
  | ‾‾14 cwt. 1 qr. 4·5 lb.

**42.**

1. £·225
2. £·975
3. 2·875 s.
4. ·4375 gu.
5. £1·05416̇
6. £·7302083̇
7. ·275 lb. av.
8. ·16̇ oz. tr.
9. 3·142857̇ cwt.
10. ·583̇ hour
11. 7·0625 ac.
12. 2·0945 cub. ft.

13. ·55 of 4·204 ac.
14. 2·75 of ·04 yd.
15. ·003̇ of 3·6 ml.
16. 4·125 of 243 ac.
17. ·075 of 3 bu. 2 pk.
18. 3·0916̇ of 1 lb. 4 oz. 10 dwt.
19. ·325 of 7 ho. 24 min.
20. ·432 of 5 cwt. 2 qr. 24 lb.
21. ·037̇ of 15·201 yd.
22. 5·24 of 3·0009 ac.
23. ·725 of 7·76 bu.
24. 3·425 of 4·003 cwt.

(5) Find the value of ·0025 ac. + 3·45 ro. + ·0076 ac. + ·009 po.

|            |   | ro. | po.       |    | ac.       |
|------------|---|-----|-----------|----|-----------|
| ·0025 ac.  | = | 0   | ″ 0·4     |    | ·0025     |
| 3·45 ro.   | = | 3   | ″18·      | or | ·8625     |
| ·0076 ac.  | = | 0   | ″ 1·216   |    | ·0076     |
| ·009 po.   | = | 0   | ″ 0·009   |    | ·00005625 |
|            |   | 3   | ″19·625   | =  | ·87265625 |

25. 2·003 ml. + ·275 ml. + 1050 yd. + ·025 ml.

26. £3·3̇ — ·5 s. + ·075 cr. — 2̇85714̇ guin.

27. ·425 ho. + ·003 min. — ·275 ho. + ·925 min.

28. Express the hectolitre, = ·343901 qr., in bu. and pk.

29. Express the Linlithgow wheat boll, = ·499128 qr., in bu. and pk.

30. Express in grains troy, a weight ·00024 lb. avoir. heavier than a kilogram, which is 2·20462 lb. avoir.

31. From Paris to Berlin by railway is a distance of 1308 kilometres, of which each is = 1093·63 yards. Express the distance in miles and yards.

32. Mercury revolves round the Sun in 87·9692580 days. Express the period of revolution in days, hours, and minutes.

33. Express in avoir. wt. the weight of a Prussian pound, which is ·46771 of 2·20486 lb. avoir.

34. Find the length in inches of the Greek foot, = $\frac{24}{24}$ of the Roman foot, which was = ·97075 foot.

35. Find the weight of 3¾ cubic feet of water at 62·455 lb. avoir. per cub. ft.

36. The radius of a circle is = ·1591549 of its circumference, which contains 360°. Find the angle whose arc is = the radius.

**43.** We call the *tenth* of a Pound Sterling, a florin. In extending the decimal division of the Pound, it was proposed to call the *hundredth* a "cent," and the *thousandth* a "mil."

$$\begin{aligned} &1 \text{ florin} &&= \text{£}{\cdot}1 = 2\text{s.} \\ &1 \text{ cent} &&= {\cdot}01 = 2\tfrac{2}{5}\text{d.} \\ &1 \text{ mil} &&= {\cdot}001 = \tfrac{24}{25}\text{ f.} \end{aligned}$$

$$1 \text{ shilling} = \tfrac{1}{2} \text{ florin}; \; 1 \text{ farthing} = \tfrac{25}{24} \text{ or } 1\tfrac{1}{24} \text{ mil.}$$

To express a sum of money as the decimal of £1, we may work as in § 41.; but to do it mentally, let us consider the following analysis:—

$$14/10\tfrac{1}{4} = \overset{\text{s.}}{14} + \overset{\text{f.}}{41} = \overset{\text{fl.}}{7} + \overset{\text{m.}}{42\tfrac{17}{24}}$$

$$\tfrac{17}{24} = \tfrac{68}{96} = \frac{70\frac{20}{24}}{100}; \; \tfrac{20}{24} = \tfrac{80}{96} = \frac{83\frac{8}{24}}{100}$$

$$14/10\tfrac{1}{4} = \text{£}{\cdot}742708\dot{3}.$$

For the *first* place, we take half the number of the shillings. For the *second* and *third* places, we express the pence and farthings as farthings, and increase the number by 1 if it is $>$ 24. For the *fourth* and *fifth* places, we multiply the number in the second and third, or when the number is $>$ 25, its excess above 25, by 4, and add 1 for every 24. For the *sixth* and *seventh* places we multiply the number in the fourth and fifth, or when the number is $>$ 25, 50, or 75, its respective excess above 25, 50, or 75, by 4, and add 1 for every 24.

When the number of shillings is *odd*, we work for the next lower even number of shillings, and add 5 to the *second* place; thus, $15/10\tfrac{1}{4} = 14/10\tfrac{1}{4} + 1/ = \text{£}{\cdot}742708\dot{3} + \text{£}{\cdot}05 = \text{£}{\cdot}792708\dot{3}$.

Since $\tfrac{1}{4}$d. $=$ £$\cdot 001041\dot{6}$, any sum of money expressed in the decimal of £1 contains no more than *six terminate* places. When there are more than six places the *seventh* is *interminate*, being either $\dot{3}$ or $\dot{6}$.

Reduce the following sums of money to the decimal of £1.

(1) $17/5\tfrac{3}{4} = \text{£}{\cdot}873958\dot{3}$.

| | s. | d. | | s. | d. | | s. | d. | | s. | d. |
|---|---|---|---|---|---|---|---|---|---|---|---|
| 1. | 12 | 6 | 7. | 14 | $5\frac{1}{2}$ | 13. | 18 | $10\frac{3}{4}$ | 19. | 7 | $0\frac{1}{4}$ |
| 2. | 18 | 3 | 8. | 16 | $3\frac{1}{2}$ | 14. | 12 | $6\frac{1}{4}$ | 20. | 13 | $5\frac{3}{4}$ |
| 3. | 4 | 9 | 9. | 12 | $4\frac{1}{4}$ | 15. | 8 | $7\frac{1}{2}$ | 21. | 14 | $2\frac{3}{4}$ |
| 4. | 2 | 7 | 10. | 6 | $4\frac{3}{4}$ | 16. | 3 | $9\frac{1}{4}$ | 22. | 19 | $1\frac{1}{4}$ |
| 5. | 12 | 10 | 11. | 8 | $3\frac{1}{4}$ | 17. | 9 | $8\frac{1}{2}$ | 23. | 3 | $8\frac{3}{4}$ |
| 6. | 14 | 8 | 12. | 18 | $11\frac{1}{2}$ | 18. | 15 | $7\frac{1}{4}$ | 24. | 19 | $11\frac{3}{4}$ |

To express a sum of money *approximately* to *three* decimal places of £1, or in florins, cents, and mils, we adopt the principle of approximate decimals (see § 38.), by increasing the

**43.** number of farthings by 1 when it is > 12, or more than half-way up to 24, and by 2 when it is > 36, or nearer to 48 than to 24; thus, 16/4¾ = £·8197916 = £·819½², being nearer to £·820 than to £·819, is approximately = £·820.

(2) Reduce 16/7¼ to *three* places of the decimal of £1.

$$16/7¼ = £·831.$$

Reduce the sums of money, Nos. 1 to 24, approximately to *three* places.

Being familiar in § 42. with the common method of finding the value of the decimal of £1, we may now consider the following plan :—

Let us find the value of £·9238. By pointing off the first place, we obtain the number of florins. Now, since 96 farthings = 1 florin, we must multiply by 96. But as 96 = 100 − 4, we put 4 times the multiplicand two places to the right, and then subtract. The number made up of the first two places on the left is the number of farthings.

£·9,238
$$£·9238 = 18/5½ \tfrac{9\,6}{1\,2\,8}. \quad \dfrac{952}{22,848}$$

(3) Find the value of £·7145.

$$£·7145 = 14/3¼ \tfrac{2\,3}{2\,3}.$$

·7,145
580
13,92

| 25. £·125 | 29. £·3125 | 33. £·4236 | 37. £·7219 |
|---|---|---|---|
| 26. ·225 | 30. ·7625 | 34. ·5168 | 38. ·8437 |
| 27. ·375 | 31. ·9875 | 35. ·8274 | 39. ·2914 |
| 28. ·975 | 32. ·5375 | 36. ·4537 | 40. ·3853 |

To obtain the value of the decimal of £1 to *the nearest farthing* without a fraction, we proceed as follows :—

Let us find the value of £·7287. We consider it approximately = £·729, which is = 14 s. + 29 mils.

Since 25 mils = 24 f., we subtract 1 from 29, and obtain 29 mils = 28 f. nearly, and £·729 approximately = 14/7.

To obtain the number of shillings, we divide the number of *cents* in the first *two* places by 5, the number of mils being = the remainder with the figure in the third place annexed ; thus, £·883 = 17s. + 33 mils ; ·824 = 16s. + 24 mils. When the second figure is < 5, we may obtain the number of shillings by doubling the figure in the first place.

In reducing the number of mils to farthings, we adopt the principle of approximate decimals, and subtract 1 when the number is > 12, or more than half-way up to 25, and 2 when > 37 or nearer to 50 than to 25.

**43.** (4) Express £·768 to *the nearest farthing*.

£·768 = 15 s. + 18 m. = 15/4¼.

Valuate the decimals, Nos. 33 to 40, to *the nearest farthing*.

☞ The pupil may now construct a table, showing the correct and the approximate decimals of £1 from ¼d. to 1/, so that by mentally inserting the decimal for the number of shillings, the decimal of any sum may be obtained.

---

**44.** MISCELLANEOUS EXERCISES IN DECIMAL FRACTIONS.

1. Find the price of 30 Parian statuettes @ £1·775 each.

2. In January 1856, the number of days during which rain fell in Scotland was 13, and the amount which fell was 2·38 inches. Find the daily average for each of the 13 days.

3. How many ac. ro. and po. are in a park containing ·08 of 155·1875 acres?

4. If 31·75 poles are feued for £2·38125, how much is it per pole?

5. Find the sum of £·3125, ·4375s., and ·75d.

6. In March 1856, in Edinburgh, the thermometer at the highest was 51°·1, and at the lowest 29°·4. Find the difference or range.

7. Find the value of ·00375 lb. troy of sterling gold @ £3‖17‖10½ ℔ oz.

8. Of 100 parts of matter in locust beans, sugar and gum form 61·10, other vegetable matter forms 31·55, and moisture 5. Of how many parts does the remainder, which is mineral matter, consist?

9. The distance from Paris to Leipsic by railway is 1225 kilometres, each 1093·63 yards. Express it in miles.

10. Of the manure of dissolved bones ·1571 of its weight is organic matter. Find the weight of organic matter in 80 tons of manure.

11. Express the sum, ₁⁹₅ of 4⅔ + ⅓ + ⅘ of ₂⁷₃ + ₄⁴₀, as a decimal.

12. In February 1856, at Sandwick, Orkney, the barometer at the highest was 30·543 inches, and at the lowest 28·843 inches. Find the difference or range.

13. The following rents are drawn from a property:—mansion, £150·15; farm, £470·475; parks, £80·875; feus, £7·625. Express the total in £, s., d.

14. Find the price of 14 cwt. 3 qr. 14 lb. rice @ £·625 ℔ cwt.

15. The time of Jupiter's rotation on his axis is 9 ho. 55 min. 50 sec., and the period of his revolution round the sun is 4332·5848 days. Reduce the former to the decimal of the latter.

16. A line in a diagram in a book published in the sixteenth century, which now measures 6·83 inches, has shrunk to ⅔⅓ of its original length; find what it had been.

**44.** 17. A cubic inch of pure water weighs 252·458 grains, find the weight of a cylindrical inch which is ·7854 of a cubic inch.

18. A gallon of pure water weighs 10 lb. avoir.; and a cubic inch, 252·458 grains. From these data, find the content of a gallon.

19. The period of the revolution of the Earth round the Sun, measured sidereally, is 365·2563612 days, and that of Mars is 686·97964580 days. Reduce the latter to the decimal of the former.

20. The height of the Peak of Mulhacen in Spain, formerly estimated at 3555 metres, has been found to be ·156 kilometre less. Find its height in feet at 39·37079 inches ꝗ metre.

21. A gallon of pure water weighs 10 lb. avoir., find the weight in oz. of a pint of whey of which the Specific Gravity is 1·019.

☞ When we mention the *Specific Gravity* (s. g.) of a substance, we show how many times it is as heavy as pure water; thus, the s. g. of lead being 11·35, any volume of lead is 11·35 times the weight of the same volume of water whose s. g. is 1.

22. Find the weight of 12 gallons of olive oil, of which the s. g. is ·915.

23. Find the content of a block of granite 5·5 ft. long, 3·2 ft. broad, and 1·6 ft. deep.

24. A metre is = 39·37079 inches. Reduce an inch to the decimal of a metre.

25. What decimal of the whole time necessary to burn a ton of coals continuously at the same rate is that required to burn 2·20486 lb.?

26. Divide £31·4 among 6 men and 11 youths, giving a youth ·525 of a man's share.

27. The weight of a cubic foot of pure water is 999·278 oz. avoir., find the weight in lb. avoir. of the air in a room 12·5 ft. high, 16·25 ft. long, and 10·4 ft. broad, air being 815 times as light as water.

28. In March 1856, the weight of vapour in a cubic foot of air in Edinburgh was 2·24 grains. Find in the decimal of a lb. avoir. the weight of vapour in the atmosphere of a room 12 ft. in height, length, and breadth, supposing that there was no fire and that the window was open.

29. Reduce $\frac{1}{2}$ of $\frac{1}{35}$ of $\frac{3}{37}$ to a decimal.

30. Express the sum, $\frac{4}{7}$ of $1\frac{5}{8}$ + $\frac{6}{7}$ of $\frac{48}{36}$ + ·2, as a decimal.

31. Reduce $\frac{5}{8}$ guinea to the decimal of £1.

32. Express the sum of $\frac{3}{37}$ and $\frac{4}{17}$ as a decimal.

33. The Polar and Equatorial Diameters of the Earth are respectively 41,707,620, and 41,847,426 feet. Express each decimally in miles.

34. Find the number of miles in the Meridional Circumference of the Earth, supposing that it contains 40,000,000 metres, each 39·37079 inches.

**44.** 35. The Specific Gravity of Hydrogen, that of air being 1, is ·069, while that of air as compared with water is ·0012. Express the relative weight of Hydrogen as compared with water.

☞ Water is the standard for solids and liquids, and air for gases.

36. The s. g. of carbonic acid gas, that of air being 1, is 1·524. Express the relative weight of carbonic acid gas as compared with water.

37. Reduce an oz. avoir. to the decimal of an oz. troy.

38. Reduce a lb. troy to the decimal of a lb. avoir.

39. A Winchester bushel is = ·9694472 Imperial bushel. Express an Imperial bushel as the decimal of the former.

40. A zinc bar, which at 32° Fahrenheit measures 1 inch, at 212° measures 1·003 inch; find the length of a bar of the same metal at 212°, which at 32° measures 2·25 inches.

41. What decimal multiplied by $\frac{4}{5}$ of $\frac{7}{10}$ produces $\frac{8}{9}\frac{1}{2}$?

42. Divide £1·3125 equally among a number of almsmen, giving each ·375 florin. What is the number?

43. What quantity of sugar @ £·025 ⅌ lb., will cost 19·575 florins?

44. Divide the sum of ·075 and ·0075 by the difference of 7·5 and ·75.

45. The yard measure made by Bird in 1758 was 36·00023 inches long. How many times would this measure be contained in a mile.

46. In 1825, the Stirling jug or pint measure was measured in Edinburgh, and found to contain 104·2034 cubic inches. Reduce this to the decimal of an Imperial gallon.

47. On the floor of a room 10 ft. 8¼ in. long and 8·25 ft. broad, dust has accumulated to the depth of ·075 inch. Express the volume of dust in the room as the decimal of a cubic foot.

48. The maximum delivery of a reservoir is 567·07 cubic ft. of water ⅌ minute, and its minimum delivery 516·66 cubic ft. Find the number of gallons, each 277·274 cub. in., delivered on an equal average in 24 hours.

49. The mean diameter of the Earth is 7912·409 miles. Find the surface of a sphere of the same diameter, found by multiplying the square of the diameter by 3·1416.

50. Find the content of a sphere of the same diameter as the earth, found by multiplying the cube of the diameter by ·5236.

# CONTINUED FRACTIONS.

**45.** If we take a vulgar fraction, as $\frac{121}{415}$, and divide the numerator and the denominator by the numerator, we obtain $\frac{121}{415} = \frac{1}{3\frac{52}{121}}$.

Similarly, $\frac{52}{121} = \frac{1}{2\frac{17}{52}}$, and $\frac{17}{52} = \frac{1}{3\frac{1}{17}}$. We have thus $\frac{121}{415} =$

$\frac{1}{3\frac{52}{121}} = \frac{1}{3\cfrac{1}{2\frac{17}{52}}} = \frac{1}{3\cfrac{1}{2\cfrac{1}{3\frac{1}{17}}}}$   In the last form, we observe that every numerator is unity. A complex fraction, in which every numerator is *unity*, and every denominator includes the succeeding parts of the fraction, is termed a CONTINUED FRACTION.

In the foregoing process, we have obtained the continued fractions: $\frac{1}{3}$; $\cfrac{1}{3\frac{1}{2}}$; $\cfrac{1}{3\cfrac{1}{2\frac{1}{3}}}$; $\cfrac{1}{3\cfrac{1}{2\cfrac{1}{3\frac{1}{17}}}}$   These fractions are respectively $= \frac{1}{3}$, $\frac{2}{7}$, $\frac{7}{24}$, and $\frac{121}{415}$. We find that we have reproduced the original fraction $\frac{121}{415}$. As the other fractions continually approach to it in value, they are termed *Convergents*. The convergents are alternately greater and less than the original fraction.

(1) Find the convergents to $\frac{121}{415}$.

The practical method of finding the convergents is to proceed as in finding the G. C. M. of 121 and 415 (see ¿ 3.).

We may write the quotients 3, 2, 3, 17, in a column, and opposite the first we place *unity* in the Numerators' column, and the first quotient 3 in the Denominators'.

| Quot. | Num. | Den. |
|---|---|---|
| 3 | 1 | 3 |
| 2 | 2 | 7 |
| 3 | 7 | 24 |
| 17 | 121 | 415 |

In the second line the numerator is $= 2 \times 1$, and the denominator is $= 2 \times 3 + 1$.

In the third, the numerator is $= 3 \times 2 + 1$, and the denominator is $= 3 \times 7 + 3$.

In the fourth, the numerator and the denominator of the original fraction are reproduced.

The convergents are, $\frac{1}{3}$, $\frac{2}{7}$, $\frac{7}{24}$, $\frac{121}{415}$.

(2) Find the first three convergents to 3·14159.

By proceeding as in finding the G. C. M. of 14159 and 100000, we obtain the first three quotients, 7, 15, 1.

The convergents are,

| Quot. | Num. | Den. |
|---|---|---|
| 7 | 1 | 7 |
| 15 | 15 | 106 |
| 1 | 16 | 113 |

$3\frac{1}{7}$, $3\frac{15}{106}$, and $3\frac{16}{113}$; or, $\frac{22}{7}$, $\frac{333}{106}$, $\frac{355}{113}$.

Find the convergents to the following fractions :—

1. $\frac{68}{217}$. | 2. $\frac{417}{1808}$. | 3. $\frac{972}{1393}$. | 4. $\frac{3379}{4195}$.

**45.** 5. Find the convergents to $\frac{784}{1073}$.

☞ We first reduce the fraction to its lowest terms. But whether we do so or not, the fraction is reproduced in its lowest terms.

6. Find the fifth convergent to ·7854.

7. Find the third convergent to ·5236.

8. The Specific Gravity of oxygen is $\frac{1147}{1039}$ of that of carbonic acid gas. Give the fourth convergent to this fraction.

☞ Whenever a remainder is a comparatively small fraction of the corresponding divisor, the convergent obtained may be taken as a good approximation.

9. The Specific Gravity of gold is 19·35, and that of platinum is 21·47. Find the second convergent to $\frac{1935}{2147}$.

10. Venus revolves round the sun in 224·701 days, and the Earth in 365·256 days. Give the fifth and sixth convergents, which approximately show what part the former period is of the latter.

11. Mercury revolves round the sun in 87·969 days, and the Earth in 365·256 days. Give six convergents.

12. The solar year is = 365·24224 days. Give the fourth convergent.

13. A metre is = 39·37079 inches. Find the fourth convergent to the fraction which a yard is of a metre.

14. A Scotch acre is = 1·261183 Imperial acre. Find five convergents to the fraction which an Imperial is of a Scotch acre.

# PRACTICE.

PRACTICE is the method of computing by means of *Aliquot Parts*.

A number contained an exact number of times in another is an aliquot part of it: thus, 7 is an aliquot part of 21; 10/ of £1; 6/8 of £1; 14 lb. of 1 cwt.

### ALIQUOT PARTS.

| £ | | s. | d. | £ | | s. | d. | s. | | | d. |
|---|---|---|---|---|---|---|---|---|---|---|---|
| $\frac{1}{2}$ | = | 10 | ″ 0 | $\frac{1}{8}$ | = | 2 | ″ 6 | $\frac{1}{2}$ | = | | 6 |
| $\frac{1}{3}$ | = | 6 | ″ 8 | $\frac{1}{10}$ | = | 2 | ″ 0 | $\frac{1}{3}$ | = | | 4 |
| $\frac{1}{4}$ | = | 5 | ″ 0 | $\frac{1}{12}$ | = | 1 | ″ 8 | $\frac{1}{4}$ | = | | 3 |
| $\frac{1}{5}$ | = | 4 | ″ 0 | $\frac{1}{30}$ | = | 0 | ″ 8 | $\frac{1}{6}$ | = | | 2 |
| $\frac{1}{6}$ | = | 3 | ″ 4 | $\frac{1}{40}$ | = | 0 | ″ 6 | $\frac{1}{8}$ | = | | 1$\frac{1}{2}$ |

**46.** (1) Find the price of 794 yards of silk @ 2/6 ℣ yd.

| 2/6 | $\frac{1}{8}$ | £1 | | £794 | = | price of 794 yd. @ £1 ℣ yd. |
|---|---|---|---|---|---|---|
| | | | | £ 99 ″ 5 = | ″ | ″ @ 2/6 ″ |

**46.**

| | s. d. | | s. d. | | s. d. |
|---|---|---|---|---|---|
| 1. 8462 @ | 10 // 0 | 7. 8472 @ | 1 // 8 | 13. 7342 @ | 3 // 4 |
| 2. 7926 .. | 6 // 8 | 8. 7904 .. | 5 // 0 | 14. 9836 .. | 5 // 0 |
| 3. 8248 .. | 2 // 6 | 9. 8463 .. | 4 // 0 | 15. 9246 .. | 0 // 8 |
| 4. 7923 .. | 4 // 0 | 10. 9527 .. | 3 // 4 | 16. 9372 .. | 0 // 6 |
| 5. 7686 .. | 3 // 4 | 11. 3513 .. | 2 // 6 | 17. 7236 .. | 0 // 8 |
| 6. 7968 .. | 2 // 0 | 12. 6723 .. | 1 // 8 | 18. 8943 .. | 6 // 8 |

(2) What cost 7689 oranges @ $1\frac{1}{2}$d. and @ $\frac{3}{4}$d. each?

| d. | | | | |
|---|---|---|---|---|
| $1\frac{1}{2}$ | $\frac{1}{8}$ | 1/ | 7689s. . . | @ 1/ |
| | | | 961 // $1\frac{1}{2}$ . | @ $1\frac{1}{2}$d. |
| | | | £48 // 1 // $1\frac{1}{2}$ | |

| | |
|---|---|
| | 7689f. |
| | 3 |
| 4 | 23067 |
| 12 | $5766\frac{3}{4}$ |
| 2,0 | 48,0 // 6 |
| | £24 // 0 // $6\frac{3}{4}$ |

| | d. | | d. | | d. |
|---|---|---|---|---|---|
| 19. 7268 @ | 6 | 23. 8464 @ | $1\frac{1}{2}$ | 27. 6847 @ | $\frac{1}{4}$ |
| 20. 8379 .. | 4 | 24. 7932 .. | 1 | 28. 8467 .. | $\frac{3}{4}$ |
| 21. 3848 .. | 3 | 25. 7233 .. | $1\frac{1}{2}$ | 29. 6593 .. | $\frac{1}{2}$ |
| 22. 5766 .. | 2 | 26. 7923 .. | $\frac{1}{4}$ | 30. 7892 .. | $\frac{3}{4}$ |

**47.** (1) Give two aliquot parts which make up $7\frac{1}{2}$d. and $5\frac{1}{2}$d.

$$7\tfrac{1}{2}\text{d.} = \begin{cases} \text{d. } 6 = \tfrac{1}{2} \text{ of } 1/ \\ +1\tfrac{1}{2} = \tfrac{1}{4} \text{ of } 6\text{d.} \end{cases} \qquad 5\tfrac{1}{2}\text{d.} = \begin{cases} \text{d. } 4 = \tfrac{1}{3} \text{ of } 1/ \\ +1\tfrac{1}{2} = \tfrac{1}{8} \text{ of } 1/ \end{cases}$$

1. Find two aliquot parts which compose the following rates:

$3\frac{3}{4}$d.; 7d.; $7\frac{1}{4}$d.; $4\frac{1}{2}$d.; $6\frac{3}{4}$d.; $6\frac{1}{2}$d.

(2) Give two aliquot parts which make up 8/4 and 12/6.

$$8/4 = \begin{cases} 5/ = \tfrac{1}{4} \text{ of } £1 \\ +3/4 = \tfrac{1}{6} \text{ of } £1 \end{cases} \qquad 12/6 = \begin{cases} 10/ = \tfrac{1}{2} \text{ of } £1 \\ +2/6 = \tfrac{1}{4} \text{ of } 10/ \end{cases}$$

2. Find two aliquot parts which compose the following rates:

7/6; 3/9; 5/10; 6/3; 2/11; 4/8.

(3) Find the aliquot parts which, when respectively subtracted from 1/ and £1, leave 9d and 16/8.

$$9\text{d.} = \begin{cases} 1/ \\ -3\text{d.} = \tfrac{1}{4} \text{ of } 1/ \end{cases} \qquad 16/8 = \begin{cases} £1 \\ -3/4 = \tfrac{1}{6} \text{ of } £1. \end{cases}$$

3. Find aliquot parts which, when respectively subtracted from 1/ or £1, leave the following rates:

$10\frac{1}{2}$d.; 9d.; $11\frac{1}{2}$d.; 17/6; 13/4; 15/.

(4) Find three aliquot parts which make up $8\frac{1}{4}$d. and $15/7\frac{1}{2}$.

**47.**

$$8\tfrac14d. = \begin{cases} 6 = \tfrac14 \text{ of } 1/ \\ +2 = \tfrac13 \text{ of } 6d. \\ +\tfrac14 = \tfrac18 \text{ of } 2d. \end{cases} \qquad 15/7\tfrac12 = \begin{cases} 10/ = \tfrac12 \text{ of } £1 \\ + 5/ = \tfrac12 \text{ of } 10/ \\ +7\tfrac12d. = \tfrac18 \text{ of } 5/ \end{cases}$$

4. Find three aliquot parts which compose the following rates :
9¼d.; 7¼d.; 8½d.; 7/8½; 11/10½; 13/9.

**48.** (1) Find the price of 4671 loaves @ 4¾d. and @ 9d.

```
d.
4   | ⅓ | 1/  | 4671s. . @ 1/
½   | ⅛ | 4d. | 1557  . . @ 4d.
¼   | ½ | ¼d. |  194" 7½  .. ¼
                  97" 3¾  .. ¼
              2,0)184,8"11¼  @ 4¾d.
                 £92"8"11¼
```

```
d.
3 | ¼ | 1/ | 4671s. . . @ 1"0
             1167 " 9 . .. 0"3
           2,0)350,3"3 . @ 0"9
              £175"3"3
```

(2) Find the price of 846 yards of cotton @ 5½d. and
    @ 7¾d.

```
d.
4   | ⅓ | 1/ | 846s. . . @ 1/
1½  | ⅛ | 1/ | 282 . . . @ 4d.
               105"9 . .. 1½
             2,0)38,7"9 . @ 5½d.
                £19"7"9
```

```
              846d. . . @ 1d.
                8
d. | d.      | 6768 . . . @ 8d.
¼  | ¼ | 1  | 211½ . . .. ¼d
            12)6556½ . . . @ 7¾d.
            2,0)54,6"4
               £27"6"4½
```

| | d. | | | d. | | | d. |
|---|---|---|---|---|---|---|---|
| 1. 5683 @ | 9 | | 9. 6723 @ | 2¼ | | 17. 6874 @ | 2¾ |
| 2. 7324 .. | 7½ | | 10. 7247 .. | 6¾ | | 18. 8674 .. | 5¾ |
| 3. 4673 .. | 4½ | | 11. 3475 .. | 3½ | | 19. 7683 .. | |
| 4. 8423 .. | 5 | | 12. 4672 .. | 1½ | | 20. 8267 .. | |
| 5. 9675 .. | 3¾ | | 13. 2435 .. | 6¼ | | 21. 8956 .. | |
| 6. 3367 .. | 2¼ | | 14. 6724 .. | 6¼ | | 22. 5732 .. | 9¼ |
| 7. 3425 .. | 8¼ | | 15. 7233 .. | 8¼ | | 23. 746.. | 11¼ |
| 8. 8726 .. | 9¼ | | 16. 9894 .. | 11 | | 24. | 7 |

**49.** (1) Find the price of 423 yards of cloth @ 1/10, and
    @ 1/6¾.

```
                | £423 ..@£1  s.  d.
2/  | 1/10 | £1 |  42" 6 ..@2"  0
2d. | 1/12 | 2/ |  3"10"6 ..0"  2
                 £38" 5"6 @1"10
```

```
d.              s.  d.
6 | ¼ | 1/  | 423s. . . @ 1"0
¾ | ⅛ | 6d. | 211" 6 . ..0"6
              26" 5½ .. 0"0¾
            2,0)66,0"11¼  @ 1"6¾
               £33"0"11¼
```

**49.** (2) Find the price of 846 cwt. of rice @ 7/9¾, and @ 11/7½.

```
            || £846 .. @ £1   s. d.
  5/   ¼ £1 || 211″10 . @ 5″0
  2/6  ⅓ 5/ || 105″15 . ..2″6                       846s. . . @ 1/
  3¾d. ⅛ 2/6|| 13″ 4″4½ ..0″3¾         d.            11
              £330″ 9″4½ @ 7″9¾      6  ½  1/                    s. d.
                                     1½ ¼  6d.      9306 . @ 11″0
                                                    423  . .. 0″6
                                                    105″9 .. 0″1½
                                                 2,0)983,4″9 @ 11″7½
                                                    £491″14″9
```

(3) Find the amount of 793 railway fares @ 16/8, and @ 6/9.

```
           || £793 . . . @ £1″ 0″0
  3/4 ⅛ £1 || 132″ 3″4 .. 0″ 3″4                    793s. . . @ 1/
              £660″16″8 @ £0″16″8       d.           6
                                      9  ⅛  6/                 s. d.
                                                    4758 . . @ 6″0
                                                    594″9 . .. 0″9
                                                 2,0)5352″9 @ 6″9
                                                    £267″12″9
```

| | s. d. | | s. d. | | s. d. |
|---|---|---|---|---|---|
| 1. 4567 @ | 1″ 1½ | 17. 798 @ | 10″ 6 | 33. 893 @ | 16″ 8 |
| 2. 3283 .. | 1″ 7½ | 18. 742 .. | 4″ 8 | 34. 979 .. | 17″ 6 |
| 3. 5687 .. | 1″ 6¾ | 19. 467 .. | 5″ 3 | 35. 894 .. | 18″ 4 |
| 4. 8672 .. | 1″10½ | 20. 923 .. | 1″10 | 36. 897 .. | 18″ 9 |
| 5. 937 .. | 15″ 0 | 21. 916 .. | 4″ 6 | 37. 374 .. | 8″ 9 |
| 6. 423 .. | 13″ 4 | 22. 743 .. | 3″ 1½ | 38. 968 .. | 4″ 4½ |
| 7. 341 .. | 3″ 9 | 23. 123 .. | 2″ 4 | 39. 763 .. | 9″ 6 |
| 8. 876 .. | 12″ 6 | 24. 732 .. | 2″ 9¾ | 40. 423 .. | 13″ 5½ |
| 9. 827 .. | 11″ 8 | 25. 428 .. | 3″ 8 | 41. 346 .. | 9″ 9¾ |
| 10. 729 .. | 8″ 4 | 26. 293 .. | 5″ 6 | 42. 729 .. | 18″ 7½ |
| 11. 873 .. | 4″ 2 | 27. 468 .. | 2″ 3 | 43. 777 .. | 19″ 2½ |
| 12. 798 .. | 5″10 | 28. 736 .. | 10″ 8 | 44. 947 .. | 4″ 8 |
| 13. 149 .. | 10″10 | 29. 716 .. | 1″ 5½ | 45. 589 .. | 5″ 7½ |
| 14. 824 .. | 11″ 3 | 30. 637 .. | 2″ 8 | 46. 346 .. | 5″ 5 |
| 15. 899 .. | 6″ 3 | 31. 468 .. | 17″ 4 | 47. 777 .. | 9″ 9 |
| 16. 243 .. | 2″11 | 32. 823 .. | 7″ 8½ | 48. 732 .. | 7″10½ |

**50.** (1) Find the price of 783 qrs. of wheat @ { £3″11″3 ℀ qr.
                                                { £4 ″ 7″6 ..

```
         £783 . . . @ £1                       783 . . . @ £1
          3                                     5
            || 2349 . . . @£3″ 0″0                || 3915 . . . . @£5″ 0″0
  10/ ½ £1  || 391″10  .. 0″10″0       12/6 ⅛ £5 || 489″ 7″6 .. 0″12″6
  1/3 ⅛ 10/ || 48″18″9 .. 0″ 1″3                    £3425″12″6 @£4″ 7″6
              £2789″ 8″9 @£3″11″3
```

**50.** (2) Find the price of 379 quarters of barley @ £2″3″5½ ＠ qr.

379s. . . . . . . @ 1/
43
‾‾‾‾‾
1137 }. . . . . . @ 43″0
1516 }

| d. | | | | s. d. |
|---|---|---|---|---|
| 4 | ¼ | 1/ | 126″4 . . . . . ,, 0″4 |
| 1½ | ⅛ | 1/ | 47″4½ . . . . . . ,, 0″1½ |

2,0 ) 1647,0″8½ . . . . ,, 43″5½
‾‾‾‾‾‾‾‾‾‾‾‾‾
£823″10″8½

| | £ s. d. | | £ s. d. | | £ s. d. |
|---|---|---|---|---|---|
| 1. 916 @ | 4″16″0 | 11. 985 @ | 7″11″8 | 21. 896 @ | 7″19″ 0 |
| 2. 169 .. | 3″15″0 | 12. 946 .. | 7″ 8″9 | 22. 846 .. | 6″ 8″ 4 |
| 3. 843 .. | 2″13″4 | 13. 853 .. | 10″16″8 | 23. 859 .. | 2″12″ 6 |
| 4. 792 .. | 9″ 3″4 | 14. 976 .. | 10″12″6 | 24. 987 ..· | 4″ 7″ 6 |
| 5. 847 .. | 3″12″0 | 15. 793 .. | 12″13″4 | 25. 739 ∴· | 4″11″ 8 |
| 6. 974 .. | 3″ 7″6 | 16. 847 .. | 11″13″9 | 26. 463 .. | 4″ 1″10½ |
| 7. 874 .. | 5″16″8 | 17. 569 .. | 6″13″4 | 27. 568 .. | 7″ 1″ 6¾ |
| 8. 734 .. | 5″ 8″4 | 18. 279 .. | 4″15″0 | 28. 984 .. | 9″ 2″ 8½ |
| 9. 986 .. | 6″15″0 | 19. 947 .. | 5″18″4 | 29. 719 .. | 25″ 9″ 8½ |
| 10. 793 .. | 7″17″6 | 20. 539 .. | 4″18″8 | 30. 346 .. | 27″15″ 6¾ |

**51.** It is often convenient to employ the FLORIN as the unit of computation.

(1) Find the price of 489 tons of coal @ 14/, and @ £1″3 ＠ ton.

489 fl. . . @ 2/          489 fl. . . . @ 2/
7                         11½
‾‾‾‾‾                      ‾‾‾‾‾
342,3 fl. . . @ 14/        244″1s.
‾‾‾‾‾‾                     5379
£342″6                    ‾‾‾‾‾
                          562,3 fl. 1s. . @ 23/
                          ‾‾‾‾‾‾
                          £562″7

The most convenient method of reducing a sum expressed in £ and s. to fl., is to annex half the number of s. to the number of £; thus, £3 ″ 4 = 32 fl.

(2) Find the price of 878 cwt. of sugar @ { £1″14″8 ＠ cwt.
                                          { £2″15″4 ...

878 fl. . . @ 2/                          878 fl. . @ 2/
17                                        28
‾‾‾‾‾                                      ‾‾‾‾‾
6146 }@£1″14″0                            7024
878 }                                     1756
‾‾‾‾‾‾‾                                    ‾‾‾‾‾‾
| d. | | Fl. || 292″1″4 . .. 0″ 0″8       | 24584 . . @£2″16″0
| 8 | ⅓ | |                                | d. | | Fl. || 292″1″4 .. 0″ 0″8
1521,8″1″4 . @£1″14″8    | 8 | ⅓ |          2429,1″0″8 @£2″15″4
‾‾‾‾‾‾‾‾‾‾‾                                ‾‾‾‾‾‾‾‾‾‾
£1521″17″4                                 £2429″2″8

D 2

**51.**

| | s. | | £ s. | | £ s. d. |
|---|---|---|---|---|---|
| 1. 794 | @ 2 | 13. 943 | @ £1″16 | 25. 763 | @£1″14″ 8 |
| 2. 798 | .. 6 | 14. 879 | .. 1″18 | 26. 269 | .. 0″17″ 4 |
| 3. 823 | .. 8 | 15. 937 | .. 2″12 | 27. 263 | .. 1″13″ 6 |
| 4. 697 | .. 14 | 16. 893 | .. 3″ 4 | 28. 798 | .. 0″16″ 6 |
| 5. 796 | .. 16 | 17. 828 | .. 7″ 8 | 29. 839 | .. 0″12″ 4 |
| 6. 267 | .. 12 | 18. 726 | .. 5″14 | 30. 346 | .. 2″15″ 8 |
| 7. 937 | .. 18 | 19. 699 | .. 3″ 5 | 31. 876 | .. 0″16″ 3 |
| 8. 469 | .. 7 | 20. 893 | .. 6″ 7 | 32. 732 | .. 0″17″ 9 |
| 9. 835 | .. 11 | 21. 467 | .. 7″ 9 | 33. 356 | .. 0″14″ 2 |
| 10. 974 | .. 19 | 22. 796 | .. 9″11 | 34. 797 | .. 1″15″10 |
| 11. 826 | .. 13 | 23. 876 | .. 8″17 | 35. 798 | .. 2″18″ 2 |
| 12. 563 | .. 17 | 24. 539 | .. 11″13 | 36. 529 | .. 3″11″10 |

**52.** (1) Find the price of $749\frac{6}{11}$ cwt. @ 11/8 ℔ cwt

| | | | |
|---|---|---|---|
| | | | £749 |
| 10/ | $\frac{1}{2}$ | £1 | 374″10″0 |
| 1/8 | $\frac{1}{6}$ | 10/ | 62″ 8″4 |
| $\frac{6}{11}$ of 11/8 | | | = 0″ 6″$4\frac{1}{4}$ $\frac{5}{11}$ |
| | | | £437″ 4″$8\frac{1}{4}$ $\frac{6}{11}$ |

s. d.
11″8
6
11)70″0
6″$4\frac{1}{4}$ $\frac{5}{11}$

(2) Find the price of $292\frac{7}{16}$ lb. @ $11/5\frac{1}{2}$ ℔ lb.

| | | | |
|---|---|---|---|
| | | | £292″ 8″ 9 |
| 10/ | $\frac{1}{2}$ | £1 | 146″ 4″ $4\frac{1}{2}$ |
| 1/3 | $\frac{1}{8}$ | 10/ | 18″ 5″ $6\frac{1}{2}$ $\frac{1}{4}$ |
| $2\frac{1}{2}$d. | $\frac{1}{6}$ | 1/3 | 3″ 0″$11\frac{0}{4}$ $\frac{3}{8}$ |
| | | | £167″10″$10\frac{0}{4}$ $\frac{5}{8}$ |

Since $\frac{7}{16}$ of £1 = 8/9, the price of $292\frac{7}{16}$ lb. @ £1 is £292″8″9.

In the method of (1), we first find the price of the *integral* number 749 cwt., by taking the parts which make up the rate 11/8; and then add in the price of the *fractional* number, $\frac{6}{11}$ cwt. In the method of (2), we first find the price of the *mixed* number $292\frac{7}{16}$ lb., at the unit of computation £1, and then take the parts which make up the given rate. The first method is of more general application than the second, which is only conveniently applied when the denominator divides the unit of computation without producing a fraction.

| | s. d. | | £ s. d. | | £ s. d. |
|---|---|---|---|---|---|
| 1. $216\frac{1}{2}$ | @13″ 4 | 9. $235\frac{5}{8}$ | @0″18″ 4 | 17. $273\frac{7}{16}$ | @1 ″3 ″9 |
| 2. $547\frac{3}{4}$ | .. 16″ 8 | 10. $324\frac{3}{16}$ | .. 2″ 3″ 8 | 18. $347\frac{5}{6}$ | .. 0″17″ $5\frac{3}{4}$ |
| 3. $899\frac{1}{3}$ | .. 9″ 6 | 11. $829\frac{4}{9}$ | .. 1″ 6″ 3 | 19. $423\frac{3}{8}$ | .. 0″17″$11\frac{1}{4}$ |
| 4. $447\frac{2}{3}$ | .. 5″ 9 | 12. $247\frac{7}{20}$ | .. 1″13″ 4 | 20. $342\frac{3}{10}$ | .. 0″ 5″ $3\frac{1}{4}$ |
| 5. $967\frac{1}{4}$ | .. 6″10 | 13. $794\frac{9}{10}$ | .. 2″ 5″10 | 21. $827\frac{3}{5}$ | .. 0″19″ $2\frac{1}{2}$ |
| 6. $793\frac{3}{5}$ | .. 17″ 1 | 14. $823\frac{7}{16}$ | .. 1″ 9″ 4 | 22. $286\frac{4}{5}$ | .. 0″12″ $9\frac{1}{2}$ |
| 7. $468\frac{5}{6}$ | .. 16″ 6 | 15. $299\frac{10}{11}$ | .. 0″ 8″$11\frac{1}{4}$ | 23. $999\frac{7}{12}$ | .. 1″13″ $5\frac{1}{4}$ |
| 8. $794\frac{2}{7}$ | .. 19″ $6\frac{1}{2}$ | 16. $834\frac{5}{12}$ | .. 0″17″ 9 | 24. $889\frac{5}{8}$ | .. 2″14″ $7\frac{3}{4}$ |

## 53.

ALIQUOT PARTS.

| cwt. | | | qr. | | | ac. | | |
|---|---|---|---|---|---|---|---|---|
| $\frac{1}{2}$ | = | 2 qr. | $\frac{1}{2}$ | = | 14 lb. | $\frac{1}{2}$ | = | 2 ro. |
| $\frac{1}{4}$ | = | 1 qr. | $\frac{1}{4}$ | = | 7 lb. | $\frac{1}{4}$ | = | 1 ro. |
| $\frac{1}{7}$ | = | 16 lb. | $\frac{1}{7}$ | = | 4 lb. | $\frac{1}{5}$ | = | 32 po. |
| $\frac{1}{8}$ | = | 14 lb. | $\frac{1}{8}$ | = | $3\frac{1}{2}$ lb. | $\frac{1}{10}$ | = | 16 po. |

These Aliquot parts are given as examples. The pupil having a thorough knowledge of the Arithmetical Tables can easily find aliquot parts of the various denominations in WEIGHTS AND MEASURES.

(1) Find the price of 7 cwt. 2 qr. 7⅜ lb. @ £8″6″8 ֆ cwt.

£8 ″ 6 ″ 8 .... price of 1 cwt.

7

|  |  |  | cwt. qr. lb. |
|---|---|---|---|
|  |  | 58 ″ 6 ″ 8 .... price of 7 ″ 0 ″ 0 |
| 2 qr. | $\frac{1}{2}$ | 1 cwt. | 4 ″ 3 ″ 4 .... ... 0 ″ 2 ″ 0 |
| 7 lb. | $\frac{1}{4}$ | 2 qr. | 0 ″10 ″ 5 .... ... 0 ″ 0 ″ 7 |
| $\frac{1}{2}$ lb. | $\frac{1}{14}$ | 7 lb. | 0 ″ 0 ″ 8$\frac{3}{4}$,$\frac{5}{7}$.. ... 0 ″ 0 ″ 0$\frac{1}{2}$ |
| $\frac{1}{8}$ lb. | $\frac{1}{4}$ | $\frac{1}{2}$ lb. | 0 ″ 0 ″ 2$\frac{9}{2}$,$\frac{13}{14}$. ... 0 ″ 0 ″ 0$\frac{1}{8}$ |

£63 ″ 1 ″ 4$\frac{9}{4}$,$\frac{9}{14}$ price of 7 ″ 2 ″ 7⅜

(2) Find the rent of 353 ac. 2 ro. 10 po. @ £2″7″6 ֆ ac.

£353 ... rent of 353 ac. @ £1

2

| | | | |
|---|---|---|---|
| 5/ | $\frac{1}{4}$ | £1 | £706 ... 88 ″ 5 } .. 353 ″ 0 ″ 0 @ £2 ″ 7 ″ 6 |
| 2/6 | $\frac{1}{2}$ | 5/ | 44 ″ 2 ″ 6 ) |
| 2 ro. | $\frac{1}{2}$ | 1 ac. | 1 ″ 3 ″ 9 ..... 0 ″ 2 ″ 0 .. ... |
| 10 po. | $\frac{1}{4}$ | 2 ro. | 0 ″ 2 ″ 11$\frac{1}{2}\frac{1}{1}$ ... 0 ″ 0 ″ 10 .. ... |

£839 ″ 14 ″ 2$\frac{1}{2}\frac{1}{2}$ .. 353 ″ 2 ″ 10 @ £2 ″ 7 ″ 6

When the number in the name in which the price of a unit is given is small, as 7 cwt. in (1), we find its price by multiplication, and then take parts for the numbers in the lower names. But when the number is large, as 353 acres in (2), we may find its price by taking parts for the rate, and then finding the price of the numbers in the lower names as in (1).

| cwt. qr. lb. | ֆ cwt. | | yd. qr. nl. | ֆ yd. |
|---|---|---|---|---|
| 1. 13 ″ 2 ″14 @ £1″17″ 4 | | 9. 11 ″ 2 ″ 2 @ £0″15″ 4 |
| 2. 347 ″ 3 ″14 .. 1 ″19″ 8 | | 10. 19 ″ 3 ″ 3 .. 1 ″ 3 ″ 4 |
| 3. 72 ″ 2 ″16 .. 1 ″ 1 ″10 | | 11. 227 ″ 2 ″ 1 .. 2 ″ 7 ″ 6 |
| 4. 122 ″ 3 ″ 9$\frac{3}{4}$ .. 2 ″ 1 ″ 7$\frac{1}{2}$ | | 12. 313 ″ 3 ″ 3$\frac{7}{8}$ .. 3 ″11 ″ 8 |

| oz. dwt. gr. | ֆ oz. | | ac. ro. po. | ֆ ac. |
|---|---|---|---|---|
| 5. 7 ″15 ″12 @ £0 ″ 5 ″ 0 | | 13. 17 ″ 2 ″20 @ £6″10″ 0 |
| 6. 17 ″11 ″15 .. 1 ″ 1 ″ 8 | | 14. 43 ″ 3 ″35 .. 4 ″16″ 8 |
| oz. dwt. gr. | ֆ lb. | | 15. 365 ″ 1 ″19 .. 8 ″13″ 4 |
| 7. 9 ″15 ″23 @ £46 ″14″ 6 | | 16. 49 ″ 3 ″37$\frac{1}{2}$ .. 3 ″ 7 ″11 |
| 8. 11 ″14 ″22$\frac{3}{8}$ .. 37 ″ 2 ″ 6 | | |

| | qr. bu. pk. | | ℘ qr. | | gal. pt. gi. | | ℘ gal. |
|---|---|---|---|---|---|---|---|
| 17. | 7 " 4 " 2 | @ | £2 " 8 " 0 | 21. | 3 " 5 " 2 | @ | £0 " 8 " 0 |
| 18. | 11 " 7 " 2 | .. | 2 " 16 " 8 | 22. | 0 " 7 " 3 | .. | 0 " 16 " 0 |
| 19. | 0 " 7 " 3 | .. | 3 " 5 " 4 | 23. | 125 " 4 " 1 | .. | 1 " 10 " 3 |
| 20. | 301 " 5 " 1¾ | .. | 3 " 3 " 8 | 24. | 73 " 5 " 2⅔ | .. | 1 " 12 " 6 |

(3) Find the price of 195 cwt. 2 qr. 11 lb. @ £4 " 13 " 4 ℘ cwt.

£1 ℘ cwt. = 5/ ℘ qr. = 1/3 ℘ 7 lb. = 2⅐d. ℘ lb.

Since 195 cwt. 2 qr. 11 lb. = 195 cwt. + 2 qr. + 7 lb. + 4 lb., the price at £1 ℘ cwt. is = £195 + 2 × 5/ + 1/3 + 4 × 2⅐d. = £195 " 11 " 11 4/7. Having thus found the price of 195 cwt. 2 qr. 11 lb. @ £1 ℘ cwt., we proceed to find it at the required rate.

| | | | |
|---|---|---|---|
| | | | £195 " 11 " 11 4/7 |
| | | | 5 |
| | | £1 | £977 " 19 " 9 6/7 |
| 6/8 | ⅓ | | 65 " 3 " 11 6/7 |
| | | | £912 " 15 " 10 |

(4) Find the price of 14 yd. 1 qr. 2 nl. @ 6/4 ℘ yd.

1/ ℘ yd. = 3d. ℘ qr. = ¾d. ℘ nl.

The price of 14 yd. 1 qr. 2 nl. @ 1/ ℘ yd. is = 14/ + 3d. + 2 × ¾d. = 14/4½.

| | | | |
|---|---|---|---|
| | | | £0 " 14 " 4½ |
| | | | 6 |
| d. | | 1/ | £4 " 6 " 3 |
| 4 | ⅓ | | 0 " 4 " 9½ |
| | | | £4 " 11 " 0½ |

| | |
|---|---|
| £1 ℘ T. = 1/ ℘ cwt. = 3d. ℘ qr. | £1 ℘ ac. = 5/ ℘ ro. = 1½d. ℘ po. |
| £1 ℘ lb. tr. = 1/8 ℘ oz. = 1d. ℘ dwt. | £1 ℘ qr. = 2/6 ℘ bu. = 7½d. ℘ pk. |
| £1 ℘ oz. tr. = 1/ ℘ dwt. = ½d. ℘ gr. | £1 ℘ gal. = 2/6 ℘ pt. = 7½d. ℘ gi. |
| 5/ ℘ oz. tr. = 3d. ℘ dwt. = ⅛d. ℘ gr. | 1/ ℘ gal. = 1½d. ℘ pt. = ⅜d. ℘ gi. |

| | T. cwt. qr. | | ℘ T. | | ac. ro. po. | | ℘ ac. |
|---|---|---|---|---|---|---|---|
| 25. | 6 " 13 " 3 | @ | £5 " 7 " 6 | 37. | 13 " 2 " 30 | @ | £2 " 3 " 6 |
| 26. | 73 " 19 " 1 | .. | 0 " 13 " 4 | 38. | 14 " 1 " 27 | .. | 3 " 3 " 4 |
| 27. | 17 " 3 " 2 | .. | 4 " 2 " 6 | 39. | 37 " 3 " 11 | .. | 5 " 10 " 8 |

| | cwt. qr. lb. | | ℘ cwt. | | yd. qr. nl. | | ℘ yd. |
|---|---|---|---|---|---|---|---|
| 28. | 23 " 3 " 14 | @ | £1 " 8 " 4 | 40. | 7 " 2 " 3 | @ | £0 " 17 " 4 |
| 29. | 13 " 1 " 21 | .. | 2 " 10 " 0 | 41. | 8 " 3 " 1 | .. | 2 " 2 " 6 |
| 30. | 19 " 2 " 11 | .. | 0 " 14 " 6 | 42. | 23 " 2 " 2 | .. | 5 " 6 " 8 |

| | lb. oz. dwt. | | ℘ lb. | | qr. bu. pk. | | ℘ qr. |
|---|---|---|---|---|---|---|---|
| 31. | 3 " 7 " 11 | @ | £5 " 11 " 8 | 43. | 7 " 3 " 2 | @ | £1 " 3 " 1½ |
| 32. | 43 " 5 " 17 | .. | 10 " 13 " 4 | 44. | 6 " 5 " 3 | .. | 4 " 10 " 0 |
| 33. | 37 " 9 " 7 | .. | 3 " 17 " 4 | 45. | 15 " 3 " 1 | .. | 3 " 5 " 10 |

| | oz. dwt. gr. | | ℘ oz. | | gal. pt. gi. | | ℘ gal. |
|---|---|---|---|---|---|---|---|
| 34. | 7 " 13 " 17 | @ | £1 " 16 " 3 | 46. | 5 " 3 " 1 | @ | £0 " 16 " 0 |
| 35. | 6 " 17 " 9 | .. | 0 " 17 " 3 | 47. | 17 " 1 " 2 | .. | 0 " 17 " 4 |
| 36. | 3 " 5 " 13 | .. | 0 " 7 " 6 | 48. | 163 " 0 " 3 | .. | 2 " 2 " 0 |

**53.** The following special methods are useful in Avoirdupois Weight.

£1„1 ⅌ cwt. = 5/3 ⅌ qr. = 1/3$\frac{3}{4}$ ⅌ 7 lb. = 2$\frac{1}{4}$d. ⅌ lb.
1s. ⅌ cwt. = 3d. ⅌ qr. = $\frac{3}{4}$d. ⅌ 7 lb. = $\frac{3}{7}$ farthing ⅌ lb.

(5) Find the price of 4 cwt. 2 qr. 5 lb. @ £6„6s. ⅌ cwt.

| | | |
|---|---|---|
| 4 × £1„1„0 | = | £4„ 4„ 0 |
| 2 × 0„5„3 | = | 0„10„ 6 |
| 5 × 0„0„2$\frac{1}{4}$ | = | 0„ 0„11$\frac{1}{4}$ |
| | | £4„15„ 5$\frac{1}{4}$ |
| | | 6 |
| | | £28„12„ 7$\frac{1}{2}$ |

(6) Find the price of 16 cwt. 3 qr. 15 lb. @ 4/6 per cwt.

| | | |
|---|---|---|
| 16/ + 3 × 3d. | = | £0„16„ 9 |
| 2 × $\frac{3}{4}$d. + $\frac{3}{7}$f. | = | 0„ 0„ 1$\frac{1}{2}\frac{3}{7}$ |
| | | £0„16„10$\frac{1}{2}\frac{3}{7}$ |
| | | 4$\frac{1}{2}$ |
| | | £3„ 7„ 6$\frac{1}{4}\frac{6}{7}$ |
| | | 8„ 5$\frac{1}{4}\frac{3}{14}$ |
| | | £3„15„11$\frac{1}{2}\frac{1}{7}\frac{3}{4}$ |

| | cwt. | qr. | lb. | | |
|---|---|---|---|---|---|
| 49. | 3 | 3 | 12 | @ £7„7 ⅌ cwt. | |
| 50. | 27 | 1 | 18 | .. 5„5 | ... |
| 51. | 6 | 2 | 14 | @ 8/6 ⅌ cwt. | |
| 52. | 7 | 1 | 20 | .. 16/3 | ... |

We may now obtain a method for finding the price of 1 lb. when that of 1 cwt. is expressed in £ and s.

(7) Find the price ⅌ lb. @ £7„5s. per cwt.

£7 „ 5s. = £7 „ 7s. — 2s.
7 × 2$\frac{1}{4}$d. — 2 × $\frac{3}{7}$f. = 1/3$\frac{1}{2}\frac{1}{7}$ ⅌ lb.

Having given the price of 1 cwt. in £. and s., to find that of 1 lb., we multiply 2$\frac{1}{4}$d. by the *number of £.*, and $\frac{3}{7}$f. by the *difference* between the *number of £. and s.;* and *increase* or *diminish* the former product by the latter, according as the number of *s.* is > or < than that of £.

Find the price ⅌ lb. at the following rates ⅌ cwt.:—

53. £8„11s. | 54. £6„10s. | 55. £9„2s. | 56. £11„1s.

1d. ⅌ lb. = 2/4 ⅌ qr. = 9/4 ⅌ cwt.
$\frac{1}{4}$d. ⅌ lb. = 7d. ⅌ qr. = 2/4 ⅌ cwt.

(8) Find the price of 3 cwt. 1 qr. 13 lb. @ 5$\frac{1}{4}$d. ⅌ lb.

| s. d. | | £ s. d. |
|---|---|---|
| 3 × 9 „ 4 | = | 1„ 8„ 0 |
| 1 × 2 „ 4 | = | 0„ 2„ 4 |
| 13 × 0 „ 1 | = | 0„ 1„ 1 |
| | | £1„11„ 5 |
| | | 5$\frac{1}{4}$ |
| | | 7„10$\frac{1}{4}$ |
| | | 7„17„ 1 |
| | | £8„ 4„11$\frac{1}{4}$ |

(9) What cost 13 cwt. @ 2$\frac{3}{4}$d. per lb.?

| s. d. | | £ s. d. |
|---|---|---|
| 2 × 9 „ 4 | = | 0„18„ 8 |
| 3 × 2 „ 4 | = | 0„ 7„ 0 |
| | | £1„ 5„ 8 |
| | | 13 |
| | | £16„13„ 8 |

| | cwt. | qr. | lb. | | |
|---|---|---|---|---|---|
| 57. | 3 | 2 | 5 | @ 2d. ⅌ lb. | |
| 58. | 17 | 3 | 19 | .. 9$\frac{1}{4}$d. | ... |

59. 9 cwt. @ 7$\frac{1}{4}$d. ⅌ lb
60. 26 cwt. .. 4$\frac{3}{4}$d. .

**54.** Since the numbers 12 and 20 are employed in the *Money of Account*, we may easily obtain methods for finding the prices of 12 and 20 articles with some of their multiples, when that of a unit is given, which may be convenient in MENTAL COMPUTATION.

In finding the price of *One Dozen*, every *penny* in the price of the unit becomes a *shilling*. When the price of the unit is below 1/8, that of the dozen is below £1.

| | |
|---|---|
| (1) 12 @ 2d. = 2/ | (3) 12 @ 1/4 = 16/ |
| (2) 12 .. 3¼d. = 3/3 | (4) 12 .. 7/8½ = £4„12„6 |

Find the price of one dozen at the following rates ⅌ unit:—

| | | | |
|---|---|---|---|
| 1. 5d. | 3. 10¼d. | 5. 1/3 | 7. 2/8 |
| 2. 7d. | 4. 9¾d. | 6. 1/7¾ | 8. 3/5½ |

In finding the price of *Two Dozens*, every *penny* in the price of the unit becomes a *florin*. When the price of the unit is below 10d., that of the dozen is below £1.

| | |
|---|---|
| (5) 24 @ 4d. = 8/ | (7) 24 @ 1/5 = £1„14s. |
| (6) 24 .. 5¾d. = 11/6 | (8) 24 .. 2/3¼ = £2„14„6 |

Find the price of two dozens at the following rates ⅌ unit:—

| | | | |
|---|---|---|---|
| 9. 3d. | 11. 6¼d. | 13 1/6 | 15. 3/7 |
| 10. 9d. | 12. 7¾d. | 14. 2/4½ | 16. 5/8¾ |

In finding the price of *Four Dozens*, every *farthing* in the price of the unit becomes a *shilling*. When the price of the unit is below 5d., that of the four dozens is below £1.

(9) 48 @ 3d. = 12/ (10) 48 @ 1/5¼ = £3„9s.

Find the price of four dozens at the following rates ⅌ unit:—

| | | | |
|---|---|---|---|
| 17. 2d. | 19. 1½d. | 21. 7¼d. | 23. 1/6¼ |
| 18. 4d. | 20. 3¾d. | 22. 10½d. | 24. 1/10¾ |

In finding the price of *Eight Dozens*, every *farthing* in the price of the unit becomes a *florin*. When the price of the unit is below 2½d., that of the eight dozens is below £1.

(11) 96 @ 2d. = 16/. (12) 96 @ 7¼d. = £2„18s.

Find the price of eight dozens at the following rates ⅌ unit:—

25. 2¼d.; 26. 1½d.; 27. 5¾d.; 28. 1/6¾.

In finding the price of *Any Number of Dozens*, every *penny* in the price of the unit becomes as many *shillings* as there are dozens.

(13) 84 @ 4d. = £1„8s. (14) 144 @ 7¾d. = £4„13s.

Find the price of the following:—

29. 72 @ 3d. | 30. 108 @ 5½d. | 31. 144 @ 6¾d. | 32. 60 @ 8¼d.

In finding the price of *One Score*, every *shilling* in the price of the unit becomes one *pound*, and every *penny* becomes 1/8.

**54.**　　(15) 20 @ 3/6 = £3"10s.　　(16) 20 @ 7/5½ = £7"9"2.

Find the price of one score at the following rates ℛ unit:—

　　33. 7/;　34. 5/3;　35. 7¼d.;　36. 2/4½.

In finding the price of *Two Hundred and Forty* units, every *penny* in the price of the unit becomes a *pound*.

　　(17) 240 @ 5d. = £5.　　(18) 240 @ 1/2¾ = £14"15s.

Find the price of 240 units at the following rates ℛ unit:—

　　37. 8d.;　38. 7¾d.;　39. 1/11¼;　40. 5/7½.

In finding the price of *One Hundred* units, every *penny* in the price of the unit becomes 8/4, and every *shilling* becomes £5.

　　(19) 100 @ 4d. = £1"13"4.　　(20) 100 @ 5½d. = £2"5"10.

Find the price of 100 units at the following rates ℛ unit:—

　　41. 7d.;　42. 9¼d.;　43. 2/3;　44. 19/11.

---

**55.**　　MISCELLANEOUS EXERCISES IN PRACTICE.

(1) A bankrupt whose debts are £3075 offers a composition of 11/3 ℛ £. How much does he pay ?

$$
\begin{array}{|c|c|c|l}
 & & & \quad\text{£3075} \\
\hline
10/ & \tfrac{1}{2} & \text{£1} & \quad 1537"10 \\
1/3 & \tfrac{1}{8} & 10/ & \quad\ \ 192"\ 3\ "\ 9 \\
\hline
 & & & \text{£1729"13 " 9}
\end{array}
$$

(2) Find the weight of 124⅞ bushels of wheat @ 63 lb. ℛ bushel.

$$
\begin{array}{ccc}
 & & \text{cwt.}\quad\text{qr.}\quad\text{lb.} \\
 & & 124\ "\ 1\ "\ 14 \\
\hline
\begin{array}{c} \text{lb.} \\ 56 \\ 7 \end{array} & \begin{array}{c} \ \\ \tfrac{1}{2} \\ \tfrac{1}{8} \end{array} & \begin{array}{c} \ \\ 1\ \text{cwt.} \\ 56\ \text{lb.} \end{array} & \begin{array}{c} 62\ "\ 0\ "\ 21 \\ 7\ "\ 3\ "\ 2\tfrac{5}{8} \\ \hline 69\ "\ 3\ "\ 23\tfrac{5}{8} \end{array}
\end{array}
$$

Having found the weight of 124⅞ bushels @ 1 cwt. ℛ bu., we take aliquot parts for 63 lb., or 2 qr. 7 lb.

1. Find the price of 288 dressing-glasses @ 7/9 each.

2. Find the value of 840 stones of hay @ 7½d. each.

3. Find the price of 6 T. 15 cwt. oat manure @ £8"5"9 ℛ T.

4. What does a chemical manufacturer receive for 5 T. 16 cwt. 2 qr. of sulphate of ammonia @ £19"10s. ℛ T.?

5. Find the price of 17 cwt. 3 qr. 14 lb. of marine salt @ 2/6 ℛ cwt.

6. A bankrupt whose debts are £2016 offers a composition of 14/3¾ ℛ £. Find his effects.

7. How much is got for a silver epergne, weighing 7 lb. 3 oz. 10 dwt., sold second-hand @ 5/ ℛ oz. ?

8. What does an ensign receive in 365 days @ 5/3 ꝑ day?

9. A French sub-lieutenant receives 1350 francs ꝑ annum. To how much sterling is this equal, reckoning a franc at £$\frac{1}{2}\frac{1}{3}$?

10. Express in sterling the annual salary of a field-marshal of France, which is = 30,000 francs.

11. Find the value of 300 Austrian florins @ 2/0¼ each.

12. Find the value of 325 Rhenish florins @ 1/8 each.

13. Find the value of 360 Prussian dollars @ 2/10¾ each.

14. What is the value of a lac of 100,000 rupees @ 1/10¼ each?

15. What is the value in sterling of 5000 rubles @ 3/1¼ each?

16. To what sum in sterling are 1600 West Indian pistoles, each 16/, equivalent?

17. On Oct. 16, 1854, the stock of tea in London amounted to 47,522,000 lb. Find the duty @ 2/1 ꝑ lb.

18. A newspaper sold at 3¼d. has a circulation of 3500. How much is received for each issue?

19. Find the weight of 331 qr. 3 bu. of wheat @ 62 lb. ꝑ bushel.

20. Find the weight of 692 qr. 5 bu. of oats @ 42 lb. ꝑ bu.

21. Find the weight of 242 qr. 7 bu. of barley @ 54 lb. ꝑ bu.

22. What is the weight of 1248½ bu. of wheat @ 2 qr. 4 lb. ꝑ bu.?

23. Find the weight of 720¾ bu. of barley @ 1 qr. 26 lb. ꝑ bu.

24. What is the weight of 200 bu. of oats @ 1 qr. 16 lb. ꝑ bu.?

25. Find the import duty on 14 cwt. 2 qr. 14 lb. prunes @ 7/ ꝑ cwt.

26. Find the import duty on 16 cwt. 3 qr. 21 lb. Berlin wool @ 6d. ꝑ lb.

27. Find the amount of excise duty charged in England in 1855 on 83,221,004 lb. of hops @ 2d. ꝑ lb.

28. Find the amount of excise duty charged in the United Kingdom in 1855 on 166,776,234 lb. of paper @ 1½d. ꝑ lb.

29. Find the whole pay of 34 majors of Dragoon Guards and Dragoons in the British Army for 31 days, @ 19/3 each ꝑ day.

30. What did a writer's clerk whose income was £110 ꝑ annum, pay for income tax in 1855, at the rate of 11½d. ꝑ £?

31. What did a minister, whose stipend in 1856 was £326″10″5¼, pay for income tax @ 1/4 ꝑ £?

32. A bankrupt whose debts are £30,000 pays 8/3$\frac{7}{12}$ ꝑ £. How much does he pay?

33. In 1852, 590,767 oz. of gold coin were exported from the United Kingdom. Find the value @ £3″17″10½ ꝑ oz.

34. Reckoning the ducat at 4/2¼, find the value refused by *Shylock*, when he says:—

> "If every ducat in six thousand ducats
> Were in six parts, and every part a ducat,
> I would not draw them, I would have my bond."

(3) Find the gross rental of the following 5 farms:—

| | ac. | ro. | po. | | | £ | s. | d. |
|---|---|---|---|---|---|---|---|---|
| I. | 263 | 0 | 38 | @ | £1"11"6 . . . | 414 | 11 | $11\frac{3}{4}\ \frac{1}{10}$ |
| II. | 457 | 0 | 39 | .. | 1"5"0 . . . | 571 | 11 | $1\frac{3}{4}\ \frac{1}{2}$ |
| III. | 49 | 3 | 5 | .. | 2"5"0 . . . | 112 | 0 | $1\frac{3}{4}\ \frac{1}{2}$ |
| IV. | 156 | 2 | 32 | .. | 1"15"0 . . . | 274 | 4 | 6 |
| V. | 146 | 1 | 39 | .. | 1"13"4 . . . | 244 | 3 | $1\frac{1}{2}$ |
| | | | | | | £1616 | 10 | $10\frac{1}{4}\ \frac{1}{10}$ |

35. Find the amount of a minister's stipend:—30 qr. 7 bu. $0\frac{7}{20}$ pk. barley @ 39/6 ꝑ qr.; 12 qr. 2 bu. $3\frac{7}{20}$ pk. oats @ 24/$10\frac{1}{4}$ ꝑ qr.; 40 bolls oatmeal @ 18/10; and £48"6"$10\frac{3}{4}$.

36. In the Edinburgh grain market, on 52 Wednesdays ending Oct. 22, 1856:—42,915 qr. wheat were sold at an average price of 73/7; 42,206 qr. barley @ 42/1; 44,558 qr. oats @ 31/5. Find the amount.

37. Find the value of the average annual agricultural produce of a parish:—1386 qr. wheat @ 65/1; 1350 qr. barley @ 40/6; 2314 qr. oats @ 29/9; 82,500 stones hay @ $7\frac{1}{2}$d.; $204\frac{1}{2}$ acres turnips @ £12"2"6; $204\frac{1}{2}$ acres potatoes @ £13"1"8.

38. Find the rental of an estate containing 4 farms:—375 ac. 2 ro. 30 po. @ £3"12"6 ꝑ ac.; 432 ac. 1 ro. 20 po. @ £3"2"6 ꝑ ac.; 280 ac. 3 ro. 25 po. @ £2"12"6 ꝑ ac.; 413 ac. 0 ro. 15 po. @ £2"17 ꝑ ac.

(4)

Edinburgh,
Sept. 14, 1857.

Mrs Jones,

Bot. of Adam Coburg, General Draper,

| | | | s. d. | | £ | s. | d. |
|---|---|---|---|---|---|---|---|
| 5 pieces, each 46 yd. merino | @ | 4"3 | ꝑ yd. | | 48 | 17 | 6 |
| 8 .. .. 80 yd. cotton | .. | 0"$4\frac{1}{2}$ | .. | | 12 | 0 | 0 |
| $3\frac{1}{2}$ .. .. 54 yd. linen | .. | 2"9 | .. | | 25 | 19 | 9 |
| | | | | | £86 | 17 | 3 |

(5)

Mr James White, Grocer, Perth,

To Price & Co., Wholesale Merchants, Glasgow.

| 1857. | | | s. d. | | £ | s. | d. |
|---|---|---|---|---|---|---|---|
| June 26. | 5 chests congou, each 2 qr. 11 lb. | @ | 3"8 | ꝑ lb. | 61 | 8 | 4 |
| | 3 hhd. sugar, each 13 cwt. 2 qr. | .. | 39"4 | .. cwt. | 79 | 13 | 0 |
| Sept. 11. | 3 cwt. 1 qr. 14 lb. coffee | .. | 49"6 | .. cwt. | 8 | 7 | $0\frac{3}{4}$ |
| | 14 cwt. 2 qr. 3 lb. cheese | .. | 0"$5\frac{1}{2}$ | .. lb. | 37 | 5 | $8\frac{1}{2}$ |
| | | | | | £186 | 14 | $1\frac{1}{4}$ |

In writing out the following *Accounts*, supply *Names* and *Dates*.

39. 286 loaves @ $7\frac{1}{2}$d.; 140 loaves @ $6\frac{1}{2}$d.; 89 fancy loaves @ 8d.; 147 doz. biscuit @ 3d. per doz.; 176 lb. flour @ $2\frac{1}{2}$d.

**55.**   40. 648 silk mantles @ 14/10½; 420 richly trimmed mantles @ 45/; 600 yd. satin @ 8/10½; 252 silk velvet mantles @ 71/9; 140 Paisley shawls @ 47/6; 246 foreign shawls @ 66/8.

  41. 900 yd. moleskin @ 1/2½; 500 yd. plaiding @ 1/4; 250 yd. flannel @ 1/5; 600 yd. gingham @ 4¾d.; 1800 yd. unbleached cotton @ 3½d.; 200 yd. twilled linen @ 1/5; 80 yd. pilot cloth @ 6/5½; 200 yd. pack sheeting @ 5½d.

  42. 348 squares of Windsor soap @ 5½d. ⅌ square; 440 doz. squares of honey soap @ 10/6 ⅌ doz.; 200 bottles marrow oil @ 11¾d.; 288½ pints castor oil @ 1/2; 350 pots polishing paste @ 5¾d.; 1 cwt. 2 qr. 7 lb. starch @ 6½d. ⅌ lb.

  43. 740¾ lb. coffee, No. I., @ 1/; 370½ lb. coffee, No. II., @ 1/2; 561¼ lb. coffee, No. III., @ 1/4; 311 lb. coffee, No. IV., @ 1/8.

  44. 496⅞ qr. wheat @ 41/4; 236½ qr. barley @ 39/2; 483¾ qr. oats @ 26/1; 146¼ qr. beans @ 39/5.

  45. 14 pieces, each 37½ yd. @ 10/5 ⅌ yd.; 11 pieces, each 53½ yd. @ 12/4 ⅌ yd.; 19 pieces, each 44¾ yd. @ 13/8½ ⅌ yd.; 23 pieces, each 59¼ yd., @ 16/7¼ ⅌ yd.

  46. 124 qr. 7 bu. wheat @ 55/5 ⅌ qr.; 88 qr. 4 bu. barley @ 45/3 ⅌ qr.; 138 qr. 3 bu. oats @ 23/8 ⅌ qr.; 181 qr. 5 bu. beans @ 40/8 ⅌ qr.

  47. 6 chests congou, each 2 qr. 17 lb. @ 3/9 ⅌ lb.; 13 hhd. brown sugar, each 13 cwt. 1 qr. 18 lb. @ 36/4 ⅌ cwt.; 3 casks molasses, each 7 cwt. 2 qr. 14 lb. @ 13/9 ⅌ cwt.

  48. 14 cwt. 2 qr. 14 lb. Cheshire cheese @ 50/ ⅌ cwt.; 17 cwt. 3 qr. 14 lb. Wiltshire @ 40/ ⅌ cwt.; 23 cwt. 1 qr. 18 lb. Gouda @ 28/ ⅌ cwt.; 15 cwt. 2 qr. 13 lb. American @ 35/ ⅌ cwt.; 27 cwt. 3 qr. 16 lb. Carlow butter @ 77/ ⅌ cwt.; 39 cwt. 1 qr. 14 lb. Waterford @ 72/; 47 cwt. 2 qr. 20 lb. Dutch @ 84/ ⅌ cwt.; 23 cwt. 2 qr. 7 lb. Limerick @ 66/8 ⅌ cwt.

## 56.   ALLOWANCES ON GOODS.

IN selling goods by weight, an ALLOWANCE is made for the box or package containing them.

  The weight of any commodity, with that of the box or package containing it, is termed *Gross Weight;* the weight of the box, *Tare;* and the weight of the commodity, *Net Weight.*

  If a chest of tea weighs 80 lb., and the empty chest 16 lb., the Gross Weight is 80 lb., the Tare 16 lb., and the Net Weight 64 lb.

  *Draft* is an allowance given to a retailer to enable him to *turn the scale* in selling a commodity in small quantities.

**56.** A wholesale merchant in selling a chest of tea may deduct 1 lb. for Draft.

The Commercial Allowances *Tret* and *Cloff* are now obsolete. Cloff was similar to Draft. Tret was an allowance given on goods liable to waste.

(1) Find the net weight of 4 chests of tea, of which the gross weight and tare are respectively as follow :—

| | Gross Weight. | | | Tare. | |
|---|---|---|---|---|---|
| | cwt. | qr. | lb. | qr. | lb. |
| I. . . . | 0 ″ | 3 ″ | 10 | 0 ″ | 17 |
| II. . . . | 0 ″ | 3 ″ | 7 | 0 ″ | 16 |
| III. . . . | 0 ″ | 3 ″ | 5 | 0 ″ | 14 |
| IV. . . . | 0 ″ | 3 ″ | 4 | 0 ″ | 15 |
| | 3 ″ | 0 ″ | 26 | 2 ″ | 6 |
| | 0 ″ | 2 ″ | 6 | | |

2 ″ 2 ″ 20 Net Weight.

1. How much honey is sold, when in placing a jug weighing 7¾ oz. in one scale, weights amounting to 3 lb. 3¾ oz. are placed in the other?

2. A railway truck weighing 1 T. 16 cwt. 3 qr., when loaded with wool, weighs 8 T. 11 cwt. 1 qr. What is the weight of the wool?

3. A two-horse cart, weighing 13 cwt. 2 qr. 21 lb., when loaded with compost, weighs on the machine of a toll-bar 2 T. 1 cwt. 1 qr. 7 lb. What is the weight of the compost?

4. Find the net weight of a barrel of flour; gross weight, 1 cwt. 3 qr. 10 lb.; tare, 12 lb.

5. Find the net weight of 12 drums of Turkey figs; gross weight, 24 lb. 8 oz.; tare, ¾ lb. each.

6. Find the net weight of 3 tierces of coffee, of which the gross weight is 6 cwt. 2 qr. 9 lb. each, and the tare 2 qr. 17 lb. each.

7. Find the weight of coal brought up by a train of 20 trucks to a depôt, the average of each truck being 10 T. 17 cwt. 2 qr. gross, and 3 T. 0 cwt. 3 qr. tare.

8. Find the net weight of 3 hogsheads of sugar, of which the gross weight and the tare are as follow:

I. 13 cwt. 2 qr. 14 lb. gross; tare, 1 cwt. 1 qr.

II. 12 cwt. 1 qr. 13 lb. gross; tare, 1 cwt. 20 lb.

III. 13 cwt. 1 qr. 20 lb. gross; tare, 1 cwt. 1 qr. 7 lb.

9. Find the net weight of 3 tierces of coffee, of which the gross weight is respectively 5 cwt. 2 qr. 13 lb.; 4 cwt. 1 qr. 12 lb.; 6 cwt. 0 qr. 17 lb.; and the average tare 2 qr. 7 lb. ℣ tierce.

**56.** (2) Find the net weight of 9 bales of wool, each 3 cwt. 3 qr. 14 lb. gross; draft, 2 lb. ℈ bale; tare, 16 lb. ℈ cwt.

| cwt. | | qr. | | lb. | |
|---|---|---|---|---|---|
| 3 | ″ | 3 | ″ | 1¼ | Gross wt. of 1 bale |
| 0 | ″ | 0 | ″ | 2 | Draft ″ |
| 3 | ″ | 3 | ″ | 12 | Draft Suttle ″ |
| | | | | 9 | |

16 lb. ¼ 1 cwt.

| | cwt. | | qr. | | lb. | |
|---|---|---|---|---|---|---|
| | 31 | ″ | 2 | ″ | 2¼ | ″ ″ 9 bales |
| | 4 | ″ | 3 | ″ | 23³⁄₇ | Tare ″ |
| | 29 | ″ | 3 | ″ | 0⁴⁄₇ | Net weight ″ |

10. Find the net weight of 231 cwt. 2 qr. 3 lb. gross; tare, 14 lb. ℈ cwt.

11. Find the net weight of 200 cwt. 1 qr. 4 lb. gross; tare, 20 lb. ℈ cwt.

12. Find the net weight of 8 chests, each 1 cwt. 2 qr. 14 lb. gross; tare, 16 lb. ℈ cwt.

13. Find the net weight of 29 chests, each 1 cwt. 1 qr. 7 lb. gross; tare, 12 lb. ℈ cwt.

14. Find the net weight of five half-chests of tea, tare being 20 lb. ℈ cwt., and gross weight respectively, 1 qr. 19 lb.; 1 qr. 18 lb.; 1 qr. 20 lb.; 1 qr. 21 lb.; 1 qr. 16 lb.

15. Find the net weight of 4 chests tea, weighing respectively 75 lb., 84 lb., 63 lb., 83 lb.; draft, 1 lb. ℈ chest; tare, respectively 13 lb., 17 lb., 14 lb., 15 lb.

16. Find the net weight of 20 casks madder, average gross weight of each cask being 15 cwt. 2 qr. 14 lb.; draft, 5 lb. ℈ cask; tare, 17½ lb. ℈ cwt.

# SIMPLE PROPORTION.

In comparing two numbers, by finding how many times the one is as large as the other, the quotient obtained expresses the relation or RATIO of the dividend to the divisor; thus, the ratio of 16 to 8 is ¹⁶⁄₈; of 14 to 5, ¹⁴⁄₅; of 2 to 9, ²⁄₉.

In expressing the ratio of two numbers, as of 16 and 8, we write it thus, 16 : 8. The first term, 16, is called the *Antecedent*, and the second, 8, the *Consequent*.

Four numbers are said to be *Proportional* when the ratio of the first to the second is equal to the ratio of the third to the fourth. On examining the four numbers, 14, 8, 35, 20, we find ¹⁴⁄₈ = ³⁵⁄₂₀, or 14 : 8 = 35 : 20, and say 14 *is to* 8 *as* 35 *is to* 20, which we write as follows—14 : 8 : : 35 : 20.

Since $\frac{14}{8} = \frac{35}{20}$, $\frac{14 \times 20}{8 \times 20} = \frac{8 \times 35}{8 \times 20}$, and $14 \times 20 = 8 \times 35$. When four numbers are proportional, *the Product of the Means* is $=$ *the Product of the Extremes*.

> According to the arithmetical interpretation of Definition of Proportionality in *Euclid* (Book V. Def. 5), four numbers are proportional when the first, or a multiple of the first, contains the second as often as the third, or a like multiple of the third, contains the fourth.
>
> Let us take 8, 2, 28, 7; 8 contains 2 *four* times, and 28 contains 7 *four* times; hence 8 : 2 : : 28 : 7.
>
> Again, take 27, 48, 63, 112; *sixteen* times 27 $=$ *nine* times 48, and *sixteen* times 63 $=$ *nine* times 112; hence 27 : 48 : : 63 : 112.

In SIMPLE PROPORTION, we are required to find a number to which a given number may have a given ratio.

**57.** Find a number to which 56 may have the ratio of 24 to 63.

Let $x$ be the required number, then $24 : 63 : : 56 : x$; and since the product of the means is $=$ the product of the extremes, 24 times the required number, or $24x = 63 \times 56$, therefore the required number, $x = \frac{63 \times 56}{24} = 147$.

The fourth term in a proportion is termed the *Fourth Proportional* to the other three. We have seen it is obtained by multiplying the second term by the third, and dividing the product by the first.

(1) Find the fourth proportional to 21, 30, and 28.

$$21 : 30 : : 28 : x$$
$$x = \frac{\overset{10}{\cancel{30}} \times \overset{4}{\cancel{28}}}{\underset{7}{\cancel{21}}} = 40.$$

We may *cancel* the common factors of the first term with those of the second or the third.

Find fourth proportionals to the following numbers:

1.  6, 14, 12
2.  8, 24, 5
3.  7, 18, 21

4.  3·6, 4·2, 6·6
5.  ·27, 11·7, 2·1
6.  15·3, 2·89, ·171

(2) Find the fourth proportional to $5\frac{1}{4}$, $9\frac{4}{5}$, and $\frac{1}{4}$.

$$5\frac{1}{4} : 9\frac{4}{5} : : \frac{1}{4} : x$$
$$\frac{21}{4} : \frac{49}{5} : : \frac{1}{4} : x$$
$$x = \frac{49}{5} \times \frac{1}{4} \div \frac{21}{4} = \overset{7}{\underset{}{\frac{\cancel{49}}{5}}} \times \frac{1}{4} \times \frac{\overset{2}{\cancel{\frac{4}{}}}}{\underset{3}{\cancel{21}}} = \frac{7}{15}$$

**57.**  Find the fourth proportionals to the following numbers:

7.  $3\frac{1}{4}$, $5\frac{1}{4}$, $8\frac{1}{2}$r
8.  $3\frac{2}{9}$, $6\frac{3}{7}$, $10\frac{1}{3}$

9.  $\frac{1}{2}$, $\frac{4}{5}$, $\frac{7}{12}$
10.  $\frac{19}{27}$, $7\frac{2}{3}$, $\frac{11}{30}$

When there are three numbers, of which the first is to the second as the second is to the third, the third is termed the *Third Proportional* to the first and the second, and the second is the *Mean Proportional* between the first and the third.

(3) Find the third proportional to 9 and $11\frac{1}{4}$.

$$9 \;:\; 11\tfrac{1}{4} \;:\;:\; 11\tfrac{1}{4} \;:\; x$$

$$x = \tfrac{45}{4} \times \tfrac{45}{4} \div 9 = \frac{45 \times \overset{5}{\cancel{45}}}{4 \times 4 \times \cancel{9}} = \tfrac{225}{16} = 14\tfrac{1}{16}$$

$14\frac{1}{16}$ is the third proportional to 9 and $11\frac{1}{4}$, and $11\frac{1}{4}$ is a mean proportional between 9 and $14\frac{1}{16}$.

Find the third proportionals to the following numbers:

11.  9, 15
12.  49, 56

13.  5, 1
14.  2·75, 8·8

$\frac{1}{24}$, $\frac{1}{60}$
$\frac{8}{9}$, $\frac{7}{15}$

**58.**  (1) If 27 cwt. sugar cost £51, what cost 63 cwt.?

cwt.           £
27 . . . . . 51
63 . . . . . x

cwt.     cwt.        £        £
27  :  63  :  :  51  :  x

$$x = \frac{\overset{7}{\cancel{63}} \times \overset{17}{\cancel{51}}}{\underset{3}{\cancel{27}}} = \text{£}119.$$

The *greater* the quantity of sugar, the *greater* will be the price.  Since we multiply the third term by the second and divide by the first, in order to obtain the fourth term greater than the third, the second must be greater than the first. Having stated £51 in the third term, we place 63 cwt. in the second and 27 cwt. in the first.

Having stated the number which is of the same kind, or is homogeneous to that which is required, we place the greater, or the less of the other two homogeneous numbers, in the *second* term, according as the fourth term should be greater or less than the third.

The following method may sometimes be adopted:

Since 27 cwt. cost £51

1 cwt. costs £$\frac{51}{27}$

and 63 cwt. cost £$\frac{63 \times 51}{27}$ = £119.

**58.** (2) If 39 men can do a work in 168 days, in how many days can 72 men do it?

men.  days.
39 . . . . . 168
72 . . . . . x

men. men.  days. days.
72 : 39 : : 168 : x

$$x = \frac{\overset{7}{\cancel{168}} \times \overset{13}{\cancel{39}}}{\underset{-8-}{\cancel{72}}} = 91 \text{ days.}$$

The *greater* the number of men, the *less* will be the time. Having stated 168 days in the third term, we place 39, the less number of men, in the second term, and 72, the greater number, in the first.

We now give the following method:

Since 39 men can do a work in 168 days

1 man  ...  ...  in  $39 \times 168$ days

and 72 men  ...  ...  in  $\frac{39 \times 168}{72} = 91$ days.

In (1), since the quantity increases as the price increases, the one is said to vary *Directly* as the other. In (2), since the number of workmen increases as the number of days decreases, the one is said to vary *Inversely* as the other. The former is an example of *Direct Proportion*, the latter of *Inverse Proportion*. In Direct Proportion, the term connected with the *fourth* is always placed in the *second* term. In Inverse Proportion, the term connected with the *fourth* is always placed in the *first* term.

Every question in Proportion admits of four variations.

(1)

I. If 27 cwt. cost £51, what cost 63 cwt.?
II. If 63 cwt. cost £119, what cost 27 cwt.?
III. If 27 cwt. cost £51, how many cwt. may be had for £119?
IV. If 63 cwt. cost £119, how many cwt. may be had for £51?

(2)

I. If 39 men can do a work in 168 days, in how many days can 72 men do it?
II. If 72 men can do a work in 91 days, in how many days can 39 men do it?
III. If 72 men can do a work in 91 days, how many men can do it in 168 days?
IV. If 39 men can do a work in 168 days, how many men can do it in 91 days?

1. If 20 cwt. of rice cost £12, what cost 35 cwt.?
2. If 12 tons of linseed cake may be had for £99, how many may be had for £231?

**58.**    3. A labourer earns £35 in 40 weeks, in what time will he earn £14?

4. An express train runs 40 miles in 64 minutes; how far will it run in 24 minutes?

5. If 110 acres of a West Indian plantation can produce 200 hogsheads of sugar, find the produce of 176 acres.

6. If 48 reapers cut 20 acres in a week, how many acres will 156 reapers cut in the same time?

7. If 20 reapers can cut a field in 6 days, in what time will 30 reapers do it?

8. If 42 men can do a work in 165 days, how many men will do it in 45 days?

9. How many loaves at 8d. are equal in value to 240 loaves at 7d.?

10. A lends B £420 for 30 days; how long must B lend A £360 to return the obligation?

11. D lends E £525 for 64 days; what sum must E lend D for 48 days to return the favour?

12. If 63 oxen can be grazed in a field for 16 days, how long may 84 oxen be grazed as well in it?

13. The number of copies in the first edition of the *Lady of the Lake*, which was 2050, was to that in the second as 41 to 69. Find the number of copies in the second edition.

14. The length of the steamer track from Liverpool to Quebec, which is 2502 miles, is to that from Liverpool to Boston as 139 to 155. Find the length of the latter track.

(3) If 27 lb. of coffee cost £1 ˝ 12 ˝ 3, what cost 38 lb.?

27 lb. . . . . . . £1 ˝ 12 ˝ 3
38 lb. . . . . . .    $x$

We reduce the third term £1˝12˝3 to *pence*. We cancel the first and the third terms by 9, and obtain the fourth term in the same name as that to which the third was reduced, viz. $= \frac{38 \times 43}{3}$ *pence* $= 544\frac{2}{3}$d. $= £2˝5˝4\frac{2}{3}$.

$$
\begin{array}{ccc}
\text{lb.} & \text{lb.} & \text{£} \\
27 & : 38 & : : \quad 1˝12˝3 : x \\
\hline
3 & 43 & 20 \\
& \overline{114} & \overline{32}\text{s.} \\
& 152 & 12 \\
& \overline{3)1634} & \overline{387}\text{d.} \\
& 12)544\frac{2}{3}\text{d.} & \overline{43} \\
& 2,0)4,5\text{s. } 4\text{d.} & \\
& £2˝5˝4\frac{2}{3}\text{d.} &
\end{array}
$$

(4) If 25 yards of cloth cost £1 ˝ 14 ˝ 4¼, what cost 35 yards?

25 yd. . . . . . . £1 ˝ 14 ˝ 4¼
35 yd. . . . . . .    $x$

**58.**  We are sometimes
able to obtain the fourth
term easily without re-
ducing the third term.

$$\begin{array}{ccc} \text{yd.} & \text{yd.} & \text{£} \\ 25 : & 35 & :: 1\,''\,14\,''\,4\tfrac{1}{2} : x \\ 5 & 7 & 7 \end{array}$$

$$5)\overline{12\,''\,0\,''7\tfrac{1}{2}}$$
$$£2\,''\,8\,''1\tfrac{1}{4}$$

(5) If 4 cwt. 2 qr. 24 lb. of sugar may be had for £8″11″8,
how much sugar may be obtained for £3″15″6¼?

$$\begin{array}{lcl} 4\text{ cwt. 2 qr. 24 lb.} & \dots\dots & £8''11''8 \\ x & \dots\dots & £3''15''6\tfrac{1}{4} \end{array}$$

$$\begin{array}{ccc} £ & £ & \text{cwt. qr. lb.} \\ 8''11''8 : & 3''15''6\tfrac{1}{4} & :: 4''2''24 : x \\ 20 & 20 & 4 \end{array}$$

We reduce the first  $\overline{171}$ s.  $\overline{75}$ s.  $\overline{18}$ qr.
and the second terms to   12   12   28
*farthings,* and the third  $\overline{2060}$ d.  $\overline{906}$ d.  $\overline{148}$
term to *lb.* We cancel   4   4   38
the first and the second  $\overline{8240}$ f.  $\overline{3625}$ f.  $\overline{528}$ lb.
terms by 5, and the first  $\overline{1648}$   $\overline{725}$   $\overline{33}$
and the third by 16,    103
and obtain the fourth  $\overline{103}$   33
term $= \frac{725 \times 33}{103}$ lb.    2175
$= 232\tfrac{19}{103}$ lb.      2175

$$103)\overline{23925}(232\tfrac{29}{103}\text{ lb.}$$
$$= 2 \text{ cwt. } 0 \text{ qr. } 8\tfrac{29}{103} \text{ lb.}$$

15. If a labourer in 37 weeks saves £5″10″2¾, how much may
he save in 50 weeks?

16. Nine dozen loaves of refined sugar cost £48″7″6; what cost
73 loaves?

17. If 41 lb. of raisins may be had for £1″17″7, how many lb.
may be had for £5″11″4½?

18. If a steamer from Liverpool to Portland can make the passage of 2750 miles in 11 da. 6 ho., in what time would the passage
of 2980 miles from Liverpool to New York have likely been made?

19. In April 1857, the duty on 3 qr. 5 lb. of tea was £6″6″1.
Find the duty on 2 cwt. 1 qr. 20 lb. at the same time.

20. If a commercial traveller can drive between two towns 13
miles distant in 1 ho. 25 min., in what time can he drive 9 miles?

21. In what time will an express train, which runs at the rate of
40 miles an hour, traverse a distance which a parliamentary train,
at 24 miles an hour, runs in 3 ho. 15 min.?

22. If 1 cwt. 1 qr. 25 lb. of Mocha coffee may be had for £7″4″4½,
for what may 2 cwt. 3 qr. 11 lb. be obtained?

E

**58.** 23. If 4 cwt. 3 qr. 13 lb. of rice cost £4„10„10, how much may be bought for £3„10„8 ?

24. The annual feu-duty of a site containing 10,588 square yards is £207„11„2½. How much is it ℣ acre ?

25. If the penny loaf weighs 8 oz. avoir. when wheat is at 41/3 ℣ quarter, what should it weigh when wheat is at 49/6 ?

26. If a sum is sufficient to pay the wages of 112 workmen who get 17/6 each, how many whose wages are 24/6 may be paid with the same sum ?

27. A tierce of crushed sugar, containing 8 cwt. 3 qr. 14 lb., costs £27„10„3. What cost 7 tierces, each 7 cwt. 3 qr. 21 lb. ?

28. A box of pale soap, containing 2 cwt. 2 qr. 7 lb., costs £5„7„10½. Find the price of 7 boxes, each 2 cwt. 15 lb. ?

29. If 7 chests of tea, each 3 qr. 5 lb., cost £54„10„3, what cost 13 chests, each 3 qr. 13 lb. ?

30. If 30 yards of iron-rail weigh 17 cwt. 1 qr. 18 lb., how far will 1600 tons reach ?

(6) A bankrupt's debts are £535„10„5, and his assets £321„6„3. How much can he pay ℣ £1?

$$\text{Debts.} \qquad\qquad \text{Assets.}$$
$$£535\text{ „ }10\text{ „ }5 \quad\ldots\ldots\quad £321\text{ „ }6\text{ „ }3$$
$$£1 \qquad\qquad \ldots\ldots\qquad x$$
$$£535\text{ „ }10\text{ „ }5 \quad : \quad £1 \quad : : \quad £321\text{ „ }6\text{ „ }3 \quad : \quad x$$

Here we say *as* £535„10„5 of *debt is to* £1 of *debt, so is* £321„6„3 of *assets* to *x* of *assets.* We may however *state* and *work* as follows :—

$$£535„10„5 \ : \ £321„6„3 \ : : \ £1 \ : \ x$$

| | | |
|---|---|---|
| 20 | 20 | $\overline{20}$ s. |
| $\overline{10710}$ s. | $\overline{6426}$ s. | |
| 12 | 12 | |
| $\overline{128525}$ d. | $\overline{77115}$ d. | |
| $\overline{25705}$ | $\overline{15423}$ | |
| | 20 | |

$$25705)308460(12\text{ s.}$$
$$\underline{308460}$$

When all the terms are homogeneous, we can state the proportion in two ways.

31. A tenant whose rent is £53„6„8 pays a tax of £1„13„4. Find the tax on a rent of £36.

32. Find the rent of a tenant who pays 13/9 of poor's-rates, at the rate of 5½d. ℣ £.

**58.** 33. A bankrupt's debts are £525ʺ10ʺ6, and his assets £375ʺ7ʹ6. How much can he pay ℔ £?

34. A bankrupt whose assets are £3420 pays a composition of 9/6 ℔ £. Find the amount of his debts.

35. The tax paid on an income for the year ending April 5, 1856, was £19ʺ4, at the rate of 1/4 ℔ £. Find the income.

36. A clerk, after paying £2ʺ2ʺ1 of income-tax for the year ending April 5, 1854, found that he had £98ʺ17ʺ11 over. What was the rate ℔ £?

37. If the shadow of a staff 3 ft. 7 in. high measures 4 ft. 9 in., find the height of a steeple whose shadow is 158 ft. 4 in.

38. A farmer inadvertently used stone weights of 26 lb. 8 oz. each for 28 lb. What would 2 T. 13 cwt. of grain appear by these weights to be?

39. A merchant used weights of 27 lb. 12 oz. instead of 28 lb. Find the true weight, which would appear 1 cwt. *more* by the false weights.

(7) If a person can walk $8\frac{4}{7}$ miles in $2\frac{1}{4}$ hours, how far can he walk in $3\frac{1}{4}$ hours?

$$8\frac{4}{7} \text{ miles} \quad \ldots \quad 2\frac{1}{4} \text{ hours}$$
$$x \text{ miles} \quad \ldots \quad 3\frac{1}{4} \text{ hours}$$

$$\begin{array}{cccc} \text{ho.} & \text{ho.} & \text{ml.} & \text{ml.} \\ 2\frac{1}{4} & : & 3\frac{1}{4} & : : & 8\frac{4}{7} & : & x \end{array}$$

$$x = \frac{2}{5} \times \frac{13}{4} \times \frac{\overset{3}{\cancel{60}}}{7} = \frac{78}{7} = 11\frac{1}{7} \text{ ho.}$$

40. If $26\frac{11}{13}$ yards of cloth cost $£8\frac{13}{15}$, what cost $111\frac{1}{2}$ yards?

41. If $39\frac{9}{13}$ cwt. of rice cost $£18\frac{9}{10}$, how much rice may be had for $£3\frac{11}{14}$?

42. If $93\frac{3}{4}$ yards of damask cost $\frac{1}{4}$ of $£45\frac{3}{5}$, what cost $113\frac{7}{13}$ yards?

43. If 54 men can do a work in $29\frac{5}{8}$ days, how many men will do it in $35\frac{11}{15}$ days?

44. For every $5\frac{1}{4}$ miles that A walks, B goes $4\frac{7}{8}$ miles. How long will B take to traverse a distance walked by A in $6\frac{1}{2}$ hours?

45. A train at the rate of $25\frac{3}{8}$ miles an hour traverses a distance in $3\frac{1}{4}$ hours. In what time will one at the rate of $24\frac{1}{2}$ miles an hour traverse it?

46. A person can walk $\frac{7}{10}$ mile in $\frac{1}{11}$ of $2\frac{3}{4}$ hours. In what time can he walk $\frac{1}{4}$ of $1\frac{9}{11}$ mile at the same rate?

47. If $\frac{7}{11}$ of a vessel is worth £1393, what is the value of $\frac{7}{8}$ of $\frac{8}{11}$ of the vessel?

**58.**    (8) If 2·45 cwt. cost £22·75, how many cwt. may be had for £11·7?

$$\begin{array}{ccc} & \text{cwt.} & \text{cwt.} \\ £22\cdot75 : £11\cdot7 : : & 2\cdot45 : x \\ \overline{1\cdot75} \quad \overline{\cdot9} & \overline{49} \\ \overline{35} \quad 7 & \overline{7} \\ \overline{5} \quad 5)\overline{6\cdot3} \\ \overline{1\cdot26} \text{ cwt.} \end{array}$$

We have cancelled the first and the second terms by 13, the first and the third terms by 5. By multiplying the first and the third terms by 100, we clear the decimal points, and obtain two numbers which when cancelled by 7 are 5 and 7.

48. If 4·06 cwt. of rice cost £3·480, how much rice may be bought for £7·625?

49. A wall whose height is 9·1875 ft., casts a shadow of 10·5 ft. Find the length of the shadow of a steeple 93·8 ft. high.

50. A bar of cast-iron, whose Specific Gravity is 7·207, weighs 80 lb. Find the weight of a bar of cast-brass of the same size, whose s. g. is 8·100.

51. A jar of honey, whose s. g. is 1·450, weighs 4¾ lb. Find the weight of olive-oil, whose s. g. is ·908, contained in the same jar.

52. A block of Parian marble, whose s. g. is 2·560, weighs 2⅜ tons. Find the weight of a block of Carrara marble of the same size, whose s. g. is 2·716.

We now give some MISCELLANEOUS EXERCISES, which include several important Applications of Proportion.

53. After paying 7d. ⅌ £ as income-tax for the year ending April 5, 1854, a gentleman had £971ˮ16ˮ1 over, on what had the tax been charged?

$$\begin{array}{l} £1\text{ˮ}\ 0\text{ˮ}\ 0 \\ 0\text{ˮ}\ 0\text{ˮ}\ 7 \\ \overline{£0\text{ˮ}19\text{ˮ}5} : £971\text{ˮ}16\text{ˮ}1 : : £1 : x \end{array}$$

54. A person paid 11½d. ⅌ £ as income-tax for the year ending April 5, 1856, and had £104ˮ14ˮ7 of net proceeds. Find his income.

55. The ratio of the diameter to the circumference of a circle was given by Peter Metius as 113 : 355. Find the circumference of a fly-wheel 10 ft. in diameter.

56. A cistern can be filled by a pipe running 3⅞ gallons ⅌ minute in 54 minutes; in what time can it be filled by another running 4½ gallons ⅌ minute?

57. If 300 labourers can make an embankment in 48 days, in how many *more* days will 60 *fewer* do it?

58. 77 tailors can execute an order of regimental clothing in 30 days; how many *more* must be engaged to fulfil the order 8 days *sooner* ?

59. If 33 masons can build a wall in 47 days; and if, after working 11 days, 15 leave; in how many days *after* the 15 leave will it be finished?

   ☞ 33 masons can *finish* the wall in 47 — 11 or 36 days. Since 15 masons *have left*, 18 *remain*.

       masons.    days.

Hence, 18 : 33 : : 36 : $x$ = the number of days after the 15 have left.

60. If 17 men can do a work in 89 days; and if, after working 33 days, 3 men leave; in how many days in *all* will the work be done?

61. If 64 men can perform a work in 57 days; and if, after working for 12 days, notice is sent to finish the work 9 days *before* the stipulated time; how many *additional* men must be engaged?

62. If 3 men can do as much as 4 youths; and if 13 men can do a work in 9 days; in what time can 12 men *and* 8 youths do it?

             youths.    men.
☞         4 : 8 : : 3 : $x$ = 6
           6 + 12 = 18 men.
         men.     days.
       18 : 13 : : 9 : $x$

63. If 4 men can do as much as 7 youths; and if 15 men can do a work in 16 days; in what time can 16 men and 14 youths do it?

64. Find the Horse Power of an engine which can raise 5 tons of coals per hour from a pit whose depth is 66 fathoms.

   ☞ The labour necessary to raise 1 lb. through 1 foot is termed the *Unit of Work* (U. W.) Watt found that a horse could do 33,000 units of work ꝑ minute. 1 H. P. = 33,000 U. W.

     5 tons = 11200 lb.    66 fathoms = 396 ft.

11200 × 396 = 4435200 U. W. ꝑ ho.

6,0) 443520,0
   ‾‾‾‾‾‾‾‾
   73920 U. W. ꝑ min.    *or*  $\dfrac{11200 \times 396}{60 \times 33000}$ = no. of H. P.

 u. w.    u. w.    h. p.   h. p.
33000 : 73920 : : 1 : $x$

65. Find the H. P. of an engine which can pump 4500 gallons of water ꝑ hour from a mine whose depth is 77 fathoms.

66. A watch, set on Saturday at 8ʰ30 p. m., loses 1½ minute in 30 hours. What time does it show, next Thursday, at 4 p. m. ?

   ☞ From Saturday, 8ʰ30 p. m., to next Thursday, 4 p. m., is = 115½ hours.

   ho.    ho.    min.  min.
   30 : 115½ : : 1½ : $x$ = number of min. *before* 4.

**58.**  67. A watch, set on Friday at 9 p. m., gains 45 seconds in 12 hours.   What time does it show next Monday at 3 p. m. ?

68. A clock, set on Wednesday at 6 p. m., loses 2½ minutes daily; what is the correct time when the *clock strikes* 6 next Saturday morning ?

☞ 24 hours of *correct* time = 23 ho. 57½ min. of *clock's* time.

 ho.  min.  ho.   min. min. ⋅
 23 ″ 57½ : 60 : : 2½ : $x$ = number of min. after 6 by
         the *correct* time.

69. A sets out in a gig at the rate of 7 miles an hour.   In ¾ hour, B follows at the rate of 10 miles an hour.   In what time will B overtake A ?

☞ ¾ × 7 = 5¼ miles, the distance between A and B when B starts.

10 − 7 = 3 miles, gained by B on A every hour.

 ml.  ml.   ho.  ho.
 3 : 5¼ : : 1 : $x$ = the time in which A will be overtaken.

70. C starts from a hotel at 6 a. m., driving at the rate of 6¼ miles an hour.   At 7″45 a. m., D follows at the rate of 9⅜ miles an hour.   When will D overtake C ?

71. A luggage train starts at 5″45 a. m., at the rate of 20 miles an hour.   A parliamentary train starts from the same station at 6″20 a. m., at 25 miles an hour.   At 8″20 a. m., the luggage train shifts rails, and waits till the parliamentary train passes. When does the latter pass?

72. When do the hour and the minute hands of a watch coincide between 8 and 9 o'clock?

☞ The hour-hand moves through 5 minute-spaces while the minute-hand traverses 60.   Since the minute-hand moves 12 times as fast as the hour-hand, the former in moving through 12 spaces traverses 11 spaces more than the hour-hand.
   When the hour-hand is at VIII, the minute-hand being at XII is 40 minute-spaces behind it.   Now if the minute-hand to gain 11 spaces must move through 12, how far must it move to gain 40 spaces?

 spaces.   min.
 11 : 12 : : 40 : $x$ = number of min. after 8.

☞ The pupil may construct a table showing *all* the times when the hour and the minute hands coincide.

73. Two trains start simultaneously from the opposite termini of a railway 100 miles long: one goes at the rate of 20 miles an hour, and the other at 25 miles an hour.   When and where will they meet?

☞ The trains *approach* each other at the rate of 20 + 25 or 45 miles an hour.

**58.**

$$\begin{array}{cccc} \text{ml.} & \text{ml.} & \text{ho.} & \text{ho.} \\ 45 & : 100 & : : 1 & : x = \text{number of hours in which the trains meet.} \end{array}$$

$$\begin{array}{cccc} \text{ml.} & \text{ml.} & \text{ml.} & \text{ml.} \\ 45 & : 100 & : : 25 & : x = \text{number of miles from one of the termini.} \end{array}$$

$$100 - x = \text{number of miles from the other terminus.}$$

74. The distance from Edinburgh to Berwick by the North British Railway is 58 miles. A train starts from Edinburgh at the same time as from Berwick; the former at the rate of 24, and the latter at 30 miles ℣ hour. When and where do they meet?

75. From Carlisle to Preston is 90 miles. A train leaves Carlisle at 12ʺ15 a. m., at 40 miles ℣ hour, and Preston at 2 a. m., at 36 miles ℣ hour? When and where do they meet?

☞ Find *where* the Carlisle train is when the Preston train starts, and then proceed as in the other examples.

---

# COMPOUND PROPORTION.

**59.** WE have seen that the ratio of one number to another may be expressed by a fraction, of which the antecedent is the numerator and the consequent the denominator. Thus, the ratio of 4 to 5 is $= \frac{4}{5}$, and the ratio of 6 to 7 is $\frac{6}{7}$. Since the compound fraction $\frac{4}{5}$ of $\frac{6}{7}$ is $= \frac{24}{35}$, we say that the ratio of 24 to 35 is compounded of the ratios of 4 to 5 and of 6 to 7. Hence, if one number is to another in the ratio of 24 to 35, it is in the ratio compounded of the ratios of 4 to 5 and of 6 to 7. Thus, since $24 : 35 : : 48 : 70$, the ratio of 48 to 70 is compounded of the ratios of 4 to 5 and of 6 to 7. We write these numbers in the following form :—

$$\left.\begin{array}{c} 4 : 5 \\ 6 : 7 \end{array}\right\} : : 48 : 70.$$

$$\frac{48}{70} = \frac{4}{5} \text{ of } \frac{6}{7} = \frac{4 \times 6}{5 \times 7}.$$

$$\text{Hence } \frac{48 \times 5 \times 7}{70 \times 5 \times 7} = \frac{4 \times 6 \times 70}{5 \times 7 \times 70},$$

$$\text{and } 48 \times 5 \times 7 = 4 \times 6 \times 70.$$

*i.e.* the Product of the Means is $=$ the Product of the Extremes.

In COMPOUND PROPORTION we find a number to which a given number may have a ratio compounded of two or more ratios.

Find a number to which 72 may have a ratio compounded of the ratios $4 : 5$ and $6 : 7$.

Let $x$ be the number,

$$\left. \begin{array}{l} 4 : 5 \\ 6 : 7 \end{array} \right\} : : 72 : x.$$

Then since the product of the extremes is $=$ the product of the means,

$$4 \times 6 \times x = 72 \times 5 \times 7$$

$$\text{and } x = \frac{72 \times 5 \times 7}{4 \times 6}$$

*The required consequent is $=$ its antecedent $\times$ the other consequents $\div$ the other antecedents.*

(1) If 4 horses plough 45 acres in 10 days, in what time will 6 horses plough 81 acres?

Before stating, we may write the terms in two rows. This method is particularly useful in writing down a question to *dictation*.

4 horses,    45 acres,    10 days
6 horses,    81 acres,    $x$ days.

$$\left. \begin{array}{l} \text{Horses} \quad 6 : 4 \\ \text{Acres} \quad 45 : 81 \end{array} \right\} : : \overset{\text{Days.}}{10} : x$$

$$\overline{27,0)324,0\,(12}$$

$$or \quad x = \frac{10 \times 4 \times 81}{6 \times \cdot 45} = 12 \text{ days.}$$

We follow the same method as in Simple Proportion; thus 6 horses will take a *less* number of days than 4 horses; hence 6 : 4. Again, 81 acres will require a *greater* number of days than 45 acres; hence 45 : 81. We thus consider each pair of terms *separately* in reference to the required number.

We may work every question by resolving it into questions in Simple Proportion.

The foregoing question may be resolved as follows :—

I. If 4 horses plough 45 acres in 10 days, in what time will 6 horses plough *the same number of acres?*

Horses.        Days.

$$6 : 4 : : 10 : x = \frac{10 \times 4}{6}.$$

II. Now if 6 horses can plough 45 acres in $\frac{10 \times 4}{6}$ days, in what time will *the same number of horses* plough 81 acres?

Acres.        Days.

$$45 : 81 : : \frac{10 \times 4}{6} : x = \frac{10 \times 4 \times 81}{6 \times 45} = 12 \text{ days.}$$

**59.** (2) If 21 reapers cut 3 ac. 3 ro. of corn in $4\frac{2}{5}$ days, in what time will 24 reapers of the same strength cut 16 ac. 1 ro.?

$$\begin{array}{ccc} \text{21 reapers,} & \text{15 roods,} & \frac{2^3}{5}\text{ days} \\ \text{24 reapers,} & \text{65 roods,} & x\text{ days.} \end{array}$$

$$\begin{array}{lc} \text{Reapers} & 24:21 \\ \text{Roods} & 15:65 \end{array} \Big\} :: \frac{2^3}{5}:x.$$

$$x = \frac{23 \times \overset{7}{\cancel{21}} \times \overset{13}{\cancel{65}}}{\cancel{5} \times 24 \times \underset{5}{\cancel{15}}} = \frac{2083}{120} = 17\frac{43}{120}\text{ days.}$$

In (1) the number of days is in the *inverse* ratio of the number of horses, and in the *direct* ratio of the number of acres. In (2) the number of days is in the *inverse* ratio of the number of reapers, and in the *direct* ratio of the number of roods.

We may illustrate (1) as follows:—

4 horses plough 45 acres in 10 days

1 horse ploughs   do.   in $10 \times 4$ days

6 horses plough   do.   in $\dfrac{10 \times 4}{6}$ days

    do.   plough 1 acre   in $\dfrac{10 \times 4}{6 \times 45}$ days

    do.   plough 81 acres in $\dfrac{10 \times 4 \times 81}{6 \times 45}$ days.

Similarly we may illustrate (2) or any other exercise in Compound Proportion.

1. If 3 families of 6 persons each consume 28 loaves in a week, how many will 9 families of 5 persons each consume in the same time?

☞ This Question, and others similar to it, may easily be worked by one statement in Simple Proportion.

2. A housekeeper having used 6 pots of jelly with 14 loaves each 12 slices, wishes to know how many will be used with 8 loaves each 7 slices?

3. If 13 bushels of oats serve 3 horses for 11 days, how many bushels will serve 7 horses for 12 days?

4. If 6 boys are boarded for 10 months for £270, for what ought 13 boys to be boarded for 7 months?

5. If 8 labourers earn £14"8 in 12 days, what will 17 labourers earn in 5 days?

6. If 22,500 types are used in setting up 12 pages each 25 lines, how many types will be required in setting up 17 pages of the same type and breadth each 31 lines?

F 2

**59.** 7. A family may live for 3 months in the country for £24″10, what will be required to maintain them in town for 9 months, supposing £3 in the country to be equivalent to £4 in town?

8. If a traveller walks 140 miles in 8 days walking 7 hours a-day, how many miles may he accomplish in 12 days walking 8 hours a-day?

9. If 3 tailors make 5 vests in 11 hours, in what time will 11 tailors make 15 vests?

10. If 64 yards of carpet, 3 qr. wide, cover the floor of 4 equal rooms; how many yards of carpet, 1 yd. wide, will cover 3 of them?

11. If the 4 lb. loaf costs 8d. when wheat is @ 64/ ℔ qr., find the weight of the penny loaf when wheat is @ 56/.

12. A bootmaker who employs 15 men fulfils an order of 25 dozen pairs of Wellington boots in 4 weeks, in what time may he accomplish an order of 45 pairs by employing 3 additional men?

13. If 24 cakes can be made out of 3/ worth of oatmeal when meal is @ 18d. ℔ pk., how many cakes can be made out of 10/3¼ worth when meal is @ 13d.?

14. Captain Basil Hall, in computing the time in which Sir Walter Scott might execute the MS. of *Kenilworth*, introduces the following:—if 120 pages of 777 letters each may be written in 10 days, in what time would 3 volumes of 320 pages of 864 letters each be written?

15. A railway company charges 18/ for the carriage of 9 cwt. 40 miles,

(1) What should be charged for carrying 10 cwt. 54 miles?
(2) What weight should be carried 27 miles for 54/?
(3) How far should 3 cwt. be carried for 15/?

16. If 7 compositors set up 15 sheets in 6 days,

(1) In how many days will 21 compositors set up 30 sheets?
(2) How many sheets will 27 compositors set up in 14 days?
(3) How many compositors will set up 25 sheets in 7 days?

17. If 36 labourers clear 513 yards for a railway in 6 days,

(1) How many will clear 3800 yd. in 10 days?
(2) How many yd. will be cleared by 156 labourers in 18 days?
(3) In how many days will 16 labourers clear 190 yd.?

18. If 4 masons build 27 yards of wall in 5 days working 9 hours a-day, in how many days will 32 masons build 81 yards of a similar wall working 10 hours a-day?

19. If 12 boys are boarded 10 months for £498, find the board of 18 boys for 9 months, supposing that the cost of boarding 4 of the former = that of 3 of the latter.

20. If £5 is sufficient to maintain 8 labourers for a fortnight

**59.** when corn is at 28/ ℣ qr., how much will be required to maintain 6 labourers 29 days when corn is at 32/ ℣ qr. ?

21. If 20 men, of whom the average strength is ⅔ of an ordinary man's strength, can load 81 trucks in 8 hours; in what time will 32 men, of the average strength of ⅘ of an ordinary man's strength, load 63 trucks ?

22. If 7 labourers mow 50 acres in 9 days of 8 hours each,

(1) How many acres will 14 labourers mow in 3 days of 6 ho. ?

(2) How many labourers will mow 25 acres in 18 days of 7 ho. ?

(3) In how many days of 9 hours each will 14 labourers mow 375 acres ?

(4) By working how many hours a-day will 20 labourers mow 500 acres in 21 days ?

23. 8 men can dig a trench 200 yards long, 2 ft. broad, and 6 ft. deep, in 12 days,

(1) How many will dig another 160 yd. long, 3 ft. broad, 5 ft. deep, in 6 days ?

(2) What length of trench will 7 men dig in 11 days, supposing it 4 ft. broad and 7 ft. deep?

(3) What breadth of trench will 6 men dig in 8 days, supposing it 50 yd. long and 6 ft. deep?

(4) What depth of trench will 12 men dig in 15 days, supposing it 50 yd. long and 4 ft. broad?

24. If 17 men eat 33/ worth of bread in a week when the 4 lb. loaf is at 8d., what value of bread will 9 men eat in 2 weeks when the 2 lb. loaf is at 4½d. ?

25. If a family by using 2 gas-burners 7½ hours a-day pay £1″5 ℣ quarter when gas is @ 10/ ℣ 1000 cub. ft., what will a family using 3 burners 4 hours a-day pay ℣ quarter when gas is @ 7/6 ℣ 1000 cub. ft. ?

26. If 12 candles, of which 8 weigh 1 lb., serve 4 winter evenings from 5 to 11 P. M. ; how many candles, of which 6 weigh 1 lb., will serve 3 spring evenings from 7 to 11 P. M. ?

27. If 330 slices, $\frac{1}{14}$ inch thick, are obtained from 12 rounds of beef, how many similar rounds will supply 495 slices, ⅛ inch thick?

28. If the part representing land cut out of a map of a country 32,000 sq. miles in extent weigh 384 grains; find the extent of another country of which the part similarly cut from a map, drawn on the same kind of paper, weighs 337·5 gr., the former map being drawn on the scale of 56·25 sq. miles to the sq. in., and the latter on that of 100 sq. ml. to the sq. in.

29. If a greyhound takes 6 leaps for a hare's 7, but 4 of the hare's leaps are equal to 3 of the greyhound's, how far will the greyhound run while the hare runs 420 yards ?

**59.**  30. If the horse *Flying Dutchman* takes 10 strides while the horse *Nonsuch* takes 9, but if 6 strides of the former are equal to 5 of the latter, what distance will the latter run while the former runs 1200 yards?

31. If 6 bars of metal, 2 ft. long, 6 in. broad, and 3¼ in. thick, weigh 126 lb.; find the weight of 7 bars 3 ft. long, 4½ in. broad, and 3 in. thick.

32. The weight of 35 cubic inches of gold, of which the Specific Gravity is 19·258, is 355·270 oz. troy; find the weight of 49 cubic inches of silver, of which the Specific Gravity is 10·474.

33. A slab of granite containing 3₁⁰₆ cub. ft. weighs 541 lb., find the weight of a piece of pumice stone containing 1⅔ cub. ft., the s. g. of the former being to that of the latter as 175 to 61.

34. A contractor having engaged to lay ten miles of railway in 150 days, finds that 90 men have finished 3 miles in 80 days; how many additional men must be engaged to finish it within the time?

35. A stabler lays in 80 bushels of oats to feed 15 horses for 16 days; at the end of 4 days he receives other 5 horses; how many additional bushels will be required for the given time?

36. The diameter of the Sun is 882,000 miles. His apparent diameter as seen from the Earth is 32′ 1·8″. Find the apparent diameter of a globe of fire as large as the Solar System, 5,700,000,000 miles in diameter, viewed at the distance of the nearest fixed star 206,265 times as distant as the Sun.

———

COMPUTATIONS made at a certain rate per hundred (*per centum*) are termed PER-CENTAGES.

Per-centages are used in Commercial Arithmetic in finding Commission, Interest, &c. They are often employed in questions of Statistics.

———

# STATISTICS.

**60.** STATISTICS treats of the numerical data of any subject.

Thus, if we examine the number of persons who pay Income Tax, the amount annually paid, &c., we are said to inquire into the Statistics of the Income Tax. Again, if a Table gives the amount of Tea annually imported and consumed in Great Britain, with the amount of duty paid, &c., it is said to furnish the Statistics of the Tea Trade.

The Statistics of a country treats of its population, revenue, and general resources.

**60.**   (1) Of 93,498 births registered in Scotland in 1855, 47,872 were males. Find the per-centage.

$$93498 : 47872 : : 100 : x = 51\cdot201 \text{ per cent.}$$

1. Find the per-centage of alloy in sterling gold, of which 1 lb. troy contains 1 oz. alloy.

2. In 1851, of 335,966 emigrants from the United Kingdom, 257,372 were Irish. How much per cent. was the latter number of the whole?

3. In 1855, the produce of silver in the United Kingdom amounted to 561,300 oz., of which 4947 were from Scotland. Find the per-centage that the latter number was of the whole.

(2) The number of poor relieved in Scotland for the year ending 14th May 1848 was 100,961; for 1849, 106,434; and for 1850, 101,454. Find the increase per cent. from 1848 to 1849, and the decrease per cent. from 1849 to 1850.

$$\begin{array}{r} 106434 \\ 100961 \\ \hline \end{array}$$

$$100961 : 5473 : : 100 : x = 54\cdot2091 \text{ per cent. of increase.}$$

$$\begin{array}{r} 106434 \\ 101454 \\ \hline \end{array}$$

$$106434 : 4980 : : 100 : x = 46\cdot7896 \text{ per cent. of decrease.}$$

4. The number of letters delivered in the United Kingdom in the year preceding Dec. 5, 1839, when penny postage was generally introduced, was 82,470,596; and in 1840, 168,768,344. Find the increase per cent.

5. In 1854, the number of letters delivered in the United Kingdom was 443,649,301; and in 1855, 456,216,176. Find the increase per cent.

6. The total number of railway tickets issued in the United Kingdom in 1850 was 66,840,175; and in 1851, the year of the Great Exhibition, 78,969,623. Find the increase per cent.

7. The population of Ireland in 1841 was 8,175,124; and in 1851, 6,552,385. Find the decrease per cent.

(3) A sample of bone manure was found to contain 15·83 per cent. of sulphate of lime. Find the weight of the sulphate in 12 tons of manure.

$$\text{Tons.}$$
$$100 : 15\cdot83 : : 12 : x$$
$$\text{T.}$$
$$x = \cdot1583 \times 12 = 1\cdot9$$

**60,**      When the rate per cent. contains an *approximate* decimal, the result
can be obtained to a *certain* number of decimals only (see § 39).   In
some cases the required result is *necessarily a whole number.*

8. The number of representatives in the House of Commons
is 658.   Of this number, or even of 654, which was for some years
the number of representatives, the per-centage for Scotland is 8·1.
Find the number of the Scottish representatives.

9. A sample of turnip manure was found to contain 20·5 per cent.
of sulphate of lime.   Find the weight of the sulphate in 20 tons of
manure.

10. The Queen's Remembrancer in Scotland has a salary of
£1250 ℀ annum.   Find the salary of his chief clerk, which is 44
per cent. of his own.

(4) The Estimate for the Science and Art Department in
Scotland, for the year ending 31st March 1856, was
£1763.   Find the estimate for the succeeding year, which
gave an increase of 3·165 per cent.

$$100 : 103\text{·}165 : : £1763 : x = £1818\text{·}16$$
$$x = 1\text{·}03165 \times 1763 = £1818\text{·}8$$

11. In 1855, the number of marriages registered in Scotland was
19,639.   In 1856, the increase was at the rate of 4·318 ℀ cent.
Find the number of marriages in the latter year.

12. The population of Scotland in 1841 was 2,620,184.   Find the
population in 1851, which had increased at the rate of 10·2496
per cent.

13. The population of England and Wales in 1841 was 15,914,148.
Find the population in 1851, which had increased at the rate of
12·65202 per cent.

(5) In 1855, the per-centage of deaths, amounting to
62,154, was 2·06884 of the estimated population.   Find
the estimated population.
$$2\text{·}06884 : 100 : : 62154 : x = 3004300.$$

3,004,300 is the *reliable* number obtained from the given number
of decimal places.   Had we taken the *rate* per cent. as 2·07, we
would have obtained 3,000,000 merely.   To obtain 3,004,290, the
correctly estimated population, we require 6 decimals in the rate
per cent.

In statistical computations we can *reproduce all* the places of
*whole* numbers *only* when a *sufficient* number of decimals in the
per-centage is given.

14. In 1856, when the number of acres in Scotland on which
wheat was cultivated was 70,522 more than in 1855, the increase

**60.** was at the rate of 36·8646 per cent. Find the number on which wheat was cultivated in 1855.

15. In Scotland, during the year ending May 14, 1855, the decrease in the number of registered poor was 3217 from the former year. As the decrease was at the rate of 3·09992 per cent., find the number relieved during the year ending May 14, 1854.

(6) In December 1856, the number of deaths in London was 14,616. This was an increase of 2·482 per cent. over the number of deaths in December 1855, in which the number showed a decrease of 17·408 per cent. from that in December 1854. Find the number of deaths in December 1855 and in December 1854.

$$100 + 2\cdot482 = 102\cdot482 : 100 : : 14616 : x = 14262$$
$$100 - 17\cdot408 = 82\cdot592 : 100 : : 14262 : x = 17268$$

16. In 1851, the population of the United Kingdom, which was 27,674,352, had increased from 1841 at the rate of 73·361818 per cent. Find the population in 1841.

17. In 1812, the census of China in the seventeenth year of Kiaking amounted to 362 millions. This gave an increase of 8·7 per cent. since 1792, when a statement was made to Lord Macartney in the fifty-seventh year of Kienlung. Find the census in 1792.

18. In 1856, the number of deaths in England and Wales was 391,369; the decrease per cent. was 8·18103 from the previous year; find the number in 1855.

---

# COMMISSION AND BROKERAGE.

**61.** COMMISSION is a per-centage allowed to an agent for buying or selling goods.

BROKERAGE is a per-centage allowed to a broker for transferring the right of property, or for assisting in the sale or purchase of goods.

A merchant often allows a per-centage to a customer when he pays goods in *Ready Money*. This allowance, termed DISCOUNT, must be distinguished from *Bank Discount* (see § 64), in whose calculation the element of time is introduced.

1. Express the following per-centages as allowances ⅌ £ :

$$2\frac{1}{2} \text{ per cent.} = \frac{2\frac{1}{2}}{100} = \frac{1}{40}. \qquad \frac{1}{40} \text{ of } £1 = 6d.$$

40; 33⅓; 25; 20; 12½; 5, per cent.

61. 2. Express the following per-centages as allowances ⅌ *s.*:

25 per cent. $= \frac{25}{100} = \frac{1}{4}$. $\frac{1}{4}$ of 1s. = 3d.

50; 33⅓; 16⅔; 12½ per cent.

3. Express the following allowances as per-centages:

7/6 ⅌ £ $= \frac{7\frac{1}{2}}{20} = \frac{3}{8}$. $\frac{3}{8}$ of 100 = 37½ per cent.

10/; 5/; 2/6; 1/; 8d.; 6d. ⅌ £.
6d.; 4d.; 3d.; 2d.; 1½d.; 1d. ⅌ *s.*

(1) Find the commission on £578″10″6¼ @ 2⅜ per cent.(°/ₒ)

£100 : £578″10″6¼ : : £2⅜ : $x$

We therefore multiply the sum by the rate per cent., and divide by 100.

```
£578″10″ 6¼                          £578·526
         2⅜                               2⅜
8)1735″11″ 6¾                       8)1735578
   216″18″11¼ ⅜                        216947
  1157″ 1″ 0½                         1157052
£13,73″19″11¾ ⅜                      £13·740| = £13″14″9½
     20
   14,79      £13″14″9½ 63/160
     12
    9,59
      4
   2,39⅝
   ─────  = 315/800 = 63/160
    100
```

(2) Find the brokerage on £347″12″6 @ ⅛ °/ₒ, and @ 8/4 °/ₒ.

```
8) £347″12″6                         |£347″12″ 6
     ,43″ 9″0¾              5/  ¼ £  |  86″18″ 1½
      20                    3/4 ⅙ £  |  57″18″ 9
     8,69      8/8¼ 3/20             |£1,44″16″10½
      12                                   20
     8,28                                8,96     £1″8″11½ ½
       4                                   12
     1,15 = 15/100 = 3/20               11,62
                                            4
                                          2,5
```

**61.** Find the commission on:

| | | | |
|---|---|---|---|
| 4. £1260 | @ 5 % | 8. £375″15 | @ 3½ % |
| 5. 1274″17″8 | 4 % | 9. 509″10″6 | 4¾ % |
| 6. 375″ 7″6 | 2½ % | 10. 846″17″3 | 4⅝ % |
| 7. 840″11″6 | 5⅜ % | 11. 723″11″6 | 4/6 % |

Find the brokerage on:

| | | | |
|---|---|---|---|
| 12. £8467″10″6 | @ ⅛ % | 14. £5260″12″6 | @ 2/8 % |
| 13. 3176″13″4 | ⅜ % | 15. 324″ 3″4 | 7/3 % |

16. A commission agent sells goods to the amount of £536″10. Find his commission @ 2¼ %.

17. A broker sells 50 shares of the Bank of Scotland, each £196. Find his brokerage @ ⅛ %.

18. A traveller for a sugar-house transacted business in a provincial town to the following amount:—Raw sugar, £620; crushed sugar, £547″10; refined sugar, £320″15; molasses, £200″12″6. Find his commission @ 3 %.

19. An agent is allowed 5⅜ % for selling goods and guaranteeing the debts to his employer. His sales in a year amount to £15,375″10″6, and his losses to £375″4″2. Find his income.

20. An agent is allowed 5⅛ % for selling goods and guaranteeing the debts. His sales amount to £13,756″10″8; his bad debts to £200″15; and his doubtful debts, amounting to £500″16, are valued @ 12/6 ⅌ £. Find his probable income.

21. An agent is allowed 5¾ % for sales and risk of debts. Sales amount to £15,246″10; debts, amounting to £609″15, are valued @ 10/6 ⅌ £. Find his probable income.

22. An *invoice*, containing an account of goods purchased, is sent by an agent to his employer. The price of goods is £409″12″6; charges for packing, &c., £7″12″9; commission on the whole, @ 2¾ %. Find the amount of the invoice.

23. An agent sent to his employer in St Vincent's an *account* of the *sales* of 56 tierces of sugar, each 8 cwt. 3 qr. 16 lb. average net weight, @ 62/ ⅌ cwt.; deducting commission @ 2⅛ %; duty, 15/ ⅌ cwt.; freight, &c., £180″12″9. Find the net proceeds.

24. An agent, who is offered a commission of 5½ % on amount of sales with risk of bad debts, or a commission of 3¾ % on amount of sales without any risk, accepts the former. The sales amount to £8500, and the bad debts to £147″15. How much has he gained or lost by his choice?

# INSURANCE.

**62.** INSURANCE is a contract by which a company engages to indemnify the value of property against loss.

The owner, whose property is insured, pays to the Insurance Company a certain per-centage or *Premium* on the sum insured, on which a *Government Duty* also is chargeable. The deed of contract between the Insurance Company and the owner of the property is termed the *Policy of Insurance*.

(1) Find the expense of insuring a cargo valued at £525″12″6; premium, 2 guineas %; duty, 3/ %; commission to agent for effecting the insurance, ¼ %.

```
£525″12″6      ¼ | £525″12″6        £600 @ 3/ % = 18/
        2        |  2,62″16″3
 _____          _____         When the sum insured
 £1051″ 5″0          20             is not a multiple of £100,
   52″11″3          12,56           the duty is charged on
+ ¹⁄₁₆             _____          the next greater multiple.
 £11,03⁻16″3         12
     20            6,75
  _____           4
   ,76            _____
    12            3,
  _____
   9,15
```

```
Premium, . . . .   £11″ 0 ″ 9
Commission, . . .    2″12 ″ 6¾
Duty, . . . . .      0″18 ″ 0
                   _____
                   £14″11 ″ 3¾
```

1. Find the premium on insuring an hospital for £3400 @ 3/6 %.
2. Find the premium on insuring farm stock for £530 @ 2/6 %.
3. Find the expense of insuring household property to the amount of £469″10 @ 1/6 %; duty, 3/ %.
4. What was paid for insuring a house for £750 @ 2/6 %; duty, 3/ %?
5. What was paid for insuring a cargo for £1250 @ £1″17″6 %; duty, 2/ %?
6. An agent insures a cargo for £1370 @ 3 guineas %; duty, 4/ %; commission on the sum insured @ ½ %. What is the total expense?
7. A house factor insures four houses for £560, £940, £420, and £780 respectively, @ 1/6 %; duty, 3/ %. Find the expense.
8. Insured £3250 on a ship @ 3½ %; duty, 4/ %; commission, ½ %. Find the expense.

**62.** 9. An agent insures £4530 on a cargo @ 4½ guineas %; duty, 4/ %; commission, ½ %. Find the expense.

10. A ship, worth £5500, had a cargo worth £2670. All the expenses connected with insuring the ship and the cargo to their full value amounted to £4"1"8 %. How much was paid?

(2) Find what sum must be insured on property worth £3846, so that, in case of total loss, the whole, including the expense of insurance, may be recovered. The expense is—premium, 3 guin. %; comm$^n$, ½%; duty, 4/ %.

£3" 3
10
4

£100 — £3"17 = £96"3 : £100 : : £3846 : £4000

The expense of insuring £4000 @ £3"17 %. = £154. The net sum thus recovered = £4000 — £154 = £3846. By insuring £4000 we cover all the expenses.

11. What sum must be insured to cover £1530, the expense of insuring being £4"7"6 %?

12. How much must be insured to cover £3890; premium, 2 guin. %; comm$^n$, ½ %; duty, 3/ %?

13. How much must be insured to cover £5005; premium, £3"1 %; comm$^n$, ½ %; duty, 4/ %?

14. What sum must be insured to cover £429, all the expenses connected with the insurance being £2"10 %?

15. A cargo is worth £2442, and the expense of insuring it amounts to £2 " 17 " 6 %. What must be insured to cover the value?

# INTEREST.

**63.** INTEREST is a per-centage charged for the loan of money. The money lent is termed the *Principal*, and the sum of the Principal and the Interest is termed the *Amount*.

(1) Find the interest on £280"13"6 for 1 year @ 3¼ % ℣ annum.

£100 : £280"13"6 : : £3¼ : $x$

$$x = \frac{£280"13"6 \times 3¼}{100}.$$

## 63.

$$I = \frac{P \times R}{100}$$

| | |
|---|---|
| £280″13″6 | £280·675 |
| 3⅛ | ·035 |
| ———— | ———— |
| 140″ 6″9 | 1403375 |
| 842″ 0″6 | 842025 |
| ———— | ———— |
| £9,82″ 7″3 | £9·823625 |
| 20 | |
| ———— | |
| 16,47 | |
| 12 | |
| ———— | |
| 5,67 | £9″16″5½ ¹⁷⁄₁₅ = £9″16″5¾ *nearly.* |
| 4 | |
| ———— | |
| 2,68 | |

### Find the interest for 1 year on

1. £320        @ 3 % ℣ ann.  |  3. £802″11″6 @ 3¼ % ℣ ann.
2.   647″15″6  .. 4 %  ..   |  4.   772″16″9 .. 4¼ %  ..

(2) Find the Int. on £567″5″6 for 7 yr. @ 4½ % ℣ ann.

Prin. £100 : £567″5″6 ⎫         Int.
  Yr.  1 :      7    ⎬  : : £4½ : x
                    ⎭

$$x = \frac{£567″5″6 \times 7 \times 4½}{100}$$

Int. on a sum for a number of *years*
= Principal × N⁰ of Years × Rate % ÷ 100.

$$I = \frac{P \times Y \times R}{100}$$

| | |
|---|---|
| £567″ 5″6 | £567·275 |
| 7 | ·315 |
| ———— | ———— |
| 3970″18″6 | 1701825 |
| 4½ | 8509125 |
| ———— | ———— |
| 1985″ 9″3 | 178·691625 |
| 15883″14″0 | |
| ———— | |
| 178,69″ 3″3 | |
| 20 | |
| ———— | |
| 13,83 | £178″13″9¾ ²⁴⁄₁₅ = £178″13″10 *nearly.* |
| 12 | |
| ———— | |
| 9,99 | |
| 4 | |
| ———— | |
| 3,96 | |

**63.**

Find the Int. on

| | | | |
|---|---|---|---|
| 5. £750 for 7 yr. @ 5* % ℘ ann. | | 8. £564″13″4 for 3 yr. @ $2\frac{1}{3}$ % ℘ ann. | |
| 6. 216″ 4″6 .. 5 .. .. 3 % | | 9. 361″14″6 .. 5 .. .. $3\frac{1}{2}$ % | |
| 7. 802″17″6 .. 4 .. .. $2\frac{1}{2}$ % | | 10. 874″18″8 .. 8 .. .. $2\frac{1}{4}$ % | |

(3) Find the Int. on £321″15″$4\frac{1}{2}$ for 2 yr. 5 mo. @ $3\frac{1}{4}$ %.

| | | |
|---|---|---|
| | £321″15″$4\frac{1}{2}$ | £321·769 |
| | 2 | $2\frac{5}{12}$ |
| | 643″10″9 | 12 ) 1608845 |
| 4 mo. $\frac{1}{3}$ 1 yr. | 107″ 5″$1\frac{1}{2}$ | 134070 |
| 1 mo. $\frac{1}{4}$ 4 mo. | 26″16″$3\frac{1}{4}$ $\frac{1}{2}$ | 643538 |
| | 777″12″$1\frac{3}{4}$ $\frac{1}{2}$ | 777·608 |
| | $3\frac{1}{4}$ | $3\frac{1}{4}$ |
| | 194″ 8″$0\frac{1}{4}$ $\frac{7}{8}$ | 194402 |
| | 2332″16″$5\frac{1}{2}$ $\frac{1}{2}$ | 2332824 |
| | 100 ) 25,27″ 4″$6\frac{0}{4}$ $\frac{3}{8}$ | 100) £2527·226 |
| | £25″ 5″$5\frac{1}{4}$ $\frac{221}{800}$ | 25·27226 |
| | | = £25″5″$5\frac{1}{4}$ |

Find the Int. on

| | |
|---|---|
| 11. £374″17″3 for 5 mo. @ $3\frac{1}{3}$ % | 14. £876″14″$6\frac{1}{4}$ ..2y. 3m. @ $2\frac{1}{3}$ % |
| 12. 769″13″3 .. 8 .. .. $3\frac{3}{4}$ % | 15. 723″16″$3\frac{3}{4}$ ..3y. 11m. ... $3\frac{1}{4}$ % |
| 13. 467″ 2″$4\frac{1}{2}$ .. 5 .. .. $4\frac{1}{4}$ % | 16. 846″12″6 ..2y. 7m. ... 5 % |

(4) Find the Int. on £220″4″7 from April 1 to Sept. 11, @ 4 %.

Da.
29
31
30
31
31
11
163

Prin. £100 : £220″4″7 } : : £4 (Int.) : $x$
Da. 365 : 163 }

$$x = \frac{£220″4″7 \times 163 \times 4}{36500};$$

or, with a more convenient divisor,

$$= \frac{£220″4″7 \times 163 \times 8}{73000} = £3″18″8\tfrac{0}{4}\cdot\tfrac{1092}{1825}$$

Int. on a sum for a number of *days*

= Principal × No. of Days × Double the Rate % ÷ 73000.

---

* 5 % = 1/ ℘ £.  $2\frac{1}{2}$ % = 6d. ℘ £.

In finding Interest at the following rates, we may first take it @ 5 or $2\frac{1}{2}$ %, and then increase or diminish it as follows:—

| | | |
|---|---|---|
| 6 % = 5 % + one fifth | $5\frac{1}{2}$ % = 5 % + one tenth | 3 % = $2\frac{1}{2}$ % + one fifth |
| 4 % = 5 % — one fifth | $4\frac{1}{2}$ % = 5 % — one tenth | 2 % = $2\frac{1}{2}$ % — one fifth |

## 63.

$$I = \frac{P \times D \times 2R*}{73000}$$

By treating 73,000 in the adjoining manner, we obtain 100,000.

A number, increased in the same manner, and divided by 100,000, produces the same quotient as when divided by 73,000.

We may work (4) by this method, known as the *Third, Tenth,* and *Tenth* rule.

$$
\begin{array}{rl}
 & 73000 \\
\tfrac{1}{3} = & 24333\tfrac{1}{3} \\
\tfrac{1}{10} \text{ of } \tfrac{1}{3} = & 2433\tfrac{1}{3} \\
\tfrac{1}{10} \text{ of } \tfrac{1}{30} = & 243\tfrac{1}{3} \\
\hline
 & 100010 \\
 & -10 \\
\hline
 & 100,000
\end{array}
$$

In order to obtain the result within a farthing, we do not require the decimal in the product. The correction to be made at the end is to subtract 1 for every 10,001, or as 10,000 is sufficiently correct, we point off *four* figures, and subtract those to the left. This correction is, however, unnecessary, as in the example, when it does not affect the approximate value of the number of *mils* in the result obtained by dividing by 100,000. (See *Decimal Coinage,* § 43).

$$
\begin{array}{r}
£220{\cdot}229 \\
163 \\
\hline
660687 \\
1321374 \\
220229 \\
\hline
35897{\cdot}327 \\
8 \\
\hline
287178{\cdot}616 \\
95726 \\
9572 \\
957 \\
\hline
39,3433 \\
-39 \\
\hline
£3{\cdot}93394 \\
= £3{\prime\prime}18{\prime\prime}8\tfrac{1}{4}
\end{array}
$$

### Find the Int. on

| | | |
|---|---|---|
| 17. £420   for 73 days @ 3 % | 20. £294″18 for 231 da.@ 3% | |
| 18.  674  ..219 ..  .. 3½% | 21.  360″17  .. 120 .. ..2¼% | |
| 19.  547″10.. 88  .. .. 4¼% | 22.  301″12″6..  79 .. .. 4% | |

| | | | | | | |
|---|---|---|---|---|---|---|
| 23. £720 | from | May | 29 to July | 3 | @ 4% |
| 24.  330 | .. | June | 8 .. Sep. | 11 | .. 3% |
| 25.  690 | .. | March 10 | .. May | 29 | .. 2½% |
| 26. 2160 | .. | April | 1 .. Sep. | 11 | .. 5% |
| 27.  467″17″4 | .. | April 16 | .. June | 8 | .. 4½% |
| 28.  164″ 8″5½ | .. | Jan. | 7 .. Mar. | 29 | .. 3½% |
| 29.  876″14″6 | .. | April 2,1856, to Mar. 8, 1858, @ 3¼% | | | |
| 30.  561″ 8″3½ | .. | July 26, 1855, .. Feb.27, 1860, ..3¾% | | | |

---

\* The following may easily be verified :—

| | | | | |
|---|---|---|---|---|
| Int. for 73 days | $= \dfrac{P \times 2R}{1000}$ | | Int. for 219 days | $= \dfrac{P \times 6R}{1000}$ |
| ″ ″ 146 ″ | $= \dfrac{P \times 4R}{1000}$ | | ″ ″ 292 ″ | $= \dfrac{P \times 8R}{1000}$ |

**63.** (5) Borrowed £302·17·6 on April 1; Paid, £100 on April 29; £50·10 on June 8; and the Balance on September 11. Find the Interest due @ 3½ %.

| Dates. | Dr. | Cr. | Balances. | | Da. | Products. |
|---|---|---|---|---|---|---|
| April 1 | £302·875 | | Dr. | £302·875 | 28 | 8480·500 |
| April 29 | | £100·000 | Dr. | 202·875 | 40 | 8115·000 |
| June 8 | | 50·500 | Dr. | 152·375 | 95 | 14475·625 |
| Sept. 11 | | 152·375 | | | | 31071·125 |
| | | | | | | 7 |

73000 ) 217497·875

£2·19·7 $\frac{449}{7300}$

Sums borrowed are placed in the *Debtor* (*Dr.*) Column, and sums paid in the *Creditor* (*Cr.*) Column.

31. Borrowed £600 on June 1; Paid, £200, July 1; £300, Aug. 1. Find Int. @ 5 % due on Oct. 1.

32. Lent £950 on May 28; Received, £200, June 12; £300, July 4; Balance, Aug. 2. Find Int. @ 2½ %.

Sums *lent* are placed in *Cr.* column; sums *received* in *Dr.* column.

33. Lent £500 on Candlemas (Feb. 2); Received, £300 on Whitsunday (May 15); £100 on Lammas (Aug. 1). Find Int. @ 4¼ % due on Martinmas (Nov. 11).

34. Borrowed £525 on Lady Day (March 25); Paid, £200 on Midsummer (June 24); £150 on Michaelmas (Sep. 29). Find Int. @ 2¾ % due on Christmas (Dec. 25).

35. A barrister having borrowed £500 at the beginning of Hilary Term on Jan. 11, paid £200 at the end of Easter Term on May 8; £125 at the end of Trinity Term on June 12; and the Balance at the end of Michaelmas Term on Nov. 25. Find Int. @ 3½ %.

36. Borrowed £300·15 on Jan. 1; paid *one-fifth* on Mar. 1, and *one-fifth* on the 1st of every *second* month (May, &c.) till all was paid. Find Int. @ 5 %.

37. A capitalist advanced £3000 on Jan. 1, 1856, and received £500 on the 1st day of every quarter till the whole was paid. Find Int. @ 4 %.

38. Borrowed £506·12·6 on June 12, 1858. Paid, £200·19 on Sep. 15; £190·7·6 on Dec. 14; and £30·10 on Jan. 5, 1859. Find Int. @ 3 % due on April 5, 1859.

**63.** (6.) Borrowed £3000 on Jan. 1, 1856; £500 on Feb. 1; £1200 on March 10; £300 on July 4. Paid the whole on Aug. 2. Find Int. @ 4 %.

I.

| Dates. | Dr. | Cr. | Sums. | | Da. | Products. |
|--------|-----|-----|-------|---|-----|-----------|
| Jan. 1 | £3000 | | Dr. | £3000 | 31 | 93000 |
| Feb. 1 | 500 | | Dr. | 3500 | 38 | 133000 |
| Mar. 10 | 1200 | | Dr. | 4700 | 116 | 545200 |
| July 4 | 300 | | Dr. | 5000 | 29 | 145000 |
| Aug. 2 | | £5000 | | | | 916200 |
| | | | | | | 8 |

73000 ) 7329600

£100 8 1¼ ¼ ⅜

II.

In the second method, the *days* are reckoned to the final date: —thus from Jan. 1 to Aug. 2 = 214 da.

| Dates. | Dr. | Da | Products. |
|--------|-----|----|-----------|
| Jan. 1 | £3000 | 214 | 642000 |
| Feb. 1 | 500 | 183 | 91500 |
| Mar. 10 | 1200 | 145 | 174000 |
| July 4 | 300 | 29 | 8700 |
| Aug. 2 | | | 916200 |
| | | | 8 |

73000 ) 7329600

£100 8 1¼ ¼ ⅜

39. A freshman at Cambridge borrows 30 guineas at the beginning of Michaelmas Term, Oct. 10, 1856; 25 guineas at the beginning of Lent Term, Jan. 13, 1857; £30 at the beginning of Easter Term, April 22, 1857. Find Int. @ 4 % due at the end of Easter Term, July 10, 1857.

40. The inventor of a patent machine borrows £200 on Jan. 13, £100 on Apr. 3; £50 on May 6; £75 on July 13. Find Int. @ 4 % due on Dec. 31.

41. An Oxonian receives 50 guineas in loan on the first day of Lent, Easter, Trinity, and Michaelmas Terms, viz. Jan. 14, Apr. 22, June 3, and Oct. 10, 1857, respectively. Find Int. @ 5% due on Dec. 17, 1857.

42. Lent £509 12 6 on April 1, 1858; £392 15 6 on June 8; £96 8 6 on June 26; and £341 17 6 on Sep. 11. Find Int. @ 4½ % due on Dec. 31, 1858.

**63.** (7) Find the Interest to June 30, 1856, on the following *Account-Current*, allowing the Clydesdale Banking Company 6 %, and Mr David Deans 3¼ %.

Dr. Clydesdale Banking Co. in Acct. with Mr David Deans, Cr.

| 1856. | | | £ | s. | d. | 1856. | | £ | s. | d. |
|---|---|---|---|---|---|---|---|---|---|---|
| Jan. 10 | To Cash .. | | 310 | 0 | 0 | Feb. 14 | By Cash .. | 275 | 12 | 6 |
| Apr. 1 | " " | .. | 100 | 0 | 0 | May 12 | " " .. | 300 | 10 | 0 |
| " 29 | " " | .. | 50 | 15 | 0 | June 3 | " " .. | 50 | 13 | 6 |
| May 17 | " " | .. | 61 | 0 | 0 | " 30 | " *Balance* | 96 | 3 | 8 |
| June 24 | " " | .. | 200 | 0 | 0 | | | | | |
| " 30 | " *Interest* | | 1 | 4 | 8 | | | | | |
| | | | 722 | 19 | 8 | | | 722 | 19 | 8 |

The following shows the form of working Interest on the foregoing Account in the *Deposit Ledger* of the Bank.

| Dates. | Dr. | Cr. | Balances. | | Da. | Dr.Products. | Cr.Products. |
|---|---|---|---|---|---|---|---|
| Jan. 10 | | 310·000 | Cr. | 310·000 | 35 | | 10850·000 |
| Feb. 14 | 275·625 | | Cr. | 34·375 | 47 | | 1615·625 |
| Apr. 1 | | 100·000 | Cr. | 134·375 | 28 | | 3762·500 |
| " 29 | | 50·750 | Cr. | 185·125 | 13 | | 2406·625 |
| May 12 | 300·500 | | Dr. | 115·375 | 5 | 576·875 | |
| " 17 | | 61·000 | Dr. | 54·375 | 17 | 924·375 | |
| June 3 | 50·675 | | Dr. | 105·050 | 21 | 2206·050 | |
| " 24 | | 200·000 | Cr. | 94·950 | 6 | | 569·700 |
| " 30 | | | | | | 3707·300 | 19204·450 |
| | | | | | | 12 | · 7 |

44487·600 134431·150
44487·600
73000)89943·550

Interest due by the Bank, . . . £1ₙ4ₙ7½ ⁷⁴⁴⁴⁄₅⅛₃₅

The sums paid into the Bank are entered on the Dr. side of the pass-book, and in the Cr. column of the Bank Ledger; thus, when Mr Deans pays £310 into the Bank, the statement in the pass-book *Bank Dr. to Mr Deans* for £310, becomes in the Bank Ledger *Mr Deans Cr. by Bank* for £310. Similarly, sums drawn from the Bank are entered on the Cr. side of the pass-book, and in the Dr. column of the Bank Ledger.

The Interest on the Dr. sums in the Bank Ledger is calculated at the rate charged by the Bank, and that on the Cr. sums at the rate given by the Bank. In banks when the Dr. and Cr. Products are found, the Interest is obtained by tables; here, however, we multiply the sum of the Dr. Products by double the rate charged, and that of the Cr. Products by double the rate given, and then divide the difference of the products by 73,000. The Interest being on the Cr. side of the Bank Ledger is entered on the Dr. side of the pass-book. When the account is balanced on June 30, we find that Mr Deans has £96ₙ3ₙ8 in the Clydesdale Bank.

F

**63.**

43. Find the Int. to Dec. 31, 1855, @ 3 °/₀ on the following account of the Savings' Bank with Mr Colin Careful.

| Dr. 1855. | | 1855. | | Cr. |
|---|---|---|---|---|
| June 8 ......£15 | | Aug. 7 ......£12 | | |
| July 6 ...... 10 | | Oct. 23 ...... 8 | | |
| Sep. 5 ...... 20 | | Dec.10 ...... 10 | | |
| Nov.13 ...... 10 | | | | |

44. Find the Int. to Dec. 31, 1855, @ 2 °/₀ on the following account of the Union Bank of Scotland with Mr John Jarvie.

| Dr. 1855. | | 1855. | | Cr. |
|---|---|---|---|---|
| Mar. 10 ...£200 | | Apr. 29 ... £50 | | |
| May 29 ... 100 | | Aug. 5 ... 200 | | |
| Oct. 30 ... 300 | | | | |

45. Find the Int. to Dec. 31, 1856, @ 3½ °/₀ on the following account of the Commercial Bank of Scotland with Mr James Worthy.

| Dr. 1856. | | 1856. | | Cr. |
|---|---|---|---|---|
| July1.£155ʼʼ12ʼʼ6 | | Aug.1.£63ʼʼ12ʼʼ0 | | |
| Aug.29. 74ʼʼ15ʼʼ0 | | Oct. 1. 24ʼʼ 2ʼʼ6 | | |
| Oct. 11.100ʼʼ10ʼʼ0 | | Nov.11.26ʼʼ 5ʼʼ0 | | |
| Nov.25. 31ʼʼ17ʼʼ6 | | | | |
| Dec. 6. 42ʼʼ12ʼʼ6 | | | | |

46. Find the Int. to May 15, 1857, on the following account of the National Bank of Scotland with Mr Purdie, allowing the Bank 6°/₀, and Mr Purdie 3½ °/₀.

| Dr. 1857. | | 1857. | | Cr. |
|---|---|---|---|---|
| Jan. 6. ...£700 | | Feb. 10. ...£350 | | |
| Mar. 3. ... 120 | | Mar. 31. ... 650 | | |
| May 1. ... 200 | | May 5. ... 315 | | |
| " 11. ... 420 | | | | |

47. Find the Int. to Dec. 31, 1855, on the following account of the British Linen Company with Mr Dawson, allowing the Bank 5½°/₀, and Mr Dawson 3°/₀.

| Dr. 1855. | | 1855. | | Cr. |
|---|---|---|---|---|
| Feb. 6. ...£800 | | Mar. 5. ...£300 | | |
| Apr. 2. ... 600 | | May 31. ... 700 | | |
| July 4. ... 250 | | Aug.13. ... 850 | | |
| Oct. 9. ... 700 | | Nov.30. ... 600 | | |
| Dec. 4. ... 500 | | | | |

48. Find the Int. to June 30, 1856, on the following account of the Bank of Scotland with Mr Henderson, allowing the Bank 6 °/₀, and Mr Henderson 3½ °/₀.

| Dr. 1856. | | 1856. | | Cr. |
|---|---|---|---|---|
| Jan. 1....£1250 | | Feb. 1.... £875 | | |
| Feb.11.... 125 | | Mar. 1.... 565 | | |
| " 18.... 78 | | " 15.... 200 | | |
| Mar.29.... 231 | | Apr. 15.... 310 | | |
| May 10.... 366 | | | | |

(8) Deposited £200 in the Royal Bank of Scotland on April 10, 1855, when Interest was 3 °/₀. On May 15, Int. fell to 2½ °/₀; on June 30, to 2 °/₀; and on Oct. 8 it rose to 3 °/₀. Find the Int. due on Nov. 7.

This is an example of finding the Interest on an *Interest Receipt* for a period during which the rate varies.

| Dates. | Da. | Double Rate. | Products. |
|---|---|---|---|
| April 10 | 35 | 6 | 210 |
| May 15 | 46 | 5 | 230 |
| June 30 | 100 | 4 | 400 |
| Oct. 8 | 30 | 6 | 180 |
| Nov. 7 | | | 1020 |
| | | | 200 |

73000)204000

£2ʼʼ15ʼʼ10½ ⅔⅓

**63.** Find the Interest on the following Interest Receipts :—

49. £300 from Sep. 24 to Sep. 30, 1853, @ 2 °/₀; and to Oct. 15 @ 2½ °/₀.

50. £500 from Aug. 1 to Oct. 7, 1856, @ 2½ °/₀; to May 15, 1857, @ 3½ °/₀; and to July 10 @ 4°/₀.

51. £400 from April 1 to May 15, 1856, @ 3 °/₀; to June 30 @ 2½ °/₀, and to July 16 @ 2 °/₀.

On examining the process in (8), we see that $\dfrac{£200 \times 1020}{73000} = \dfrac{£200 \times 102 \times 2 \times 5}{73000}$
= Interest on £200 for 102 da. @ 5 °/₀.

As rates of interest may be reduced to 5 °/₀, we may consider the following plan on which Interest Tables used in some banks have been constructed.

Let a sum be deposited on May 11, when Int. is at 3 °/₀. By writing ⅗ or ·6 opposite May 12, adding ·6 continuously till the rate changes, say on May 15, to 2½ °/₀, and then adding $\frac{2\frac{1}{2}}{5}$ or ·5 continuously, we can at once see *how many days @ five °/₀* will produce the required interest.

May 11
12 ... ·6
13 ... 1·2
14 ... 1·8
15 ... 2·4
16 ... 2·9
17 ... 3·4
18 ... 3·9

Int. on £200 from May 11 to May 15 @ 3°/₀, and to May 18 @ 2½°/₀

$= \dfrac{£200 \times (4 \times 6 + 3 \times 5)}{73000} = \dfrac{£200 \times 39}{73000} = \dfrac{£200 \times 3·9 \times 10}{73000} =$

Int. on £200 @ 5 °/₀ for 3·9 days as given in the table.

(9) What Principal will produce £210 of Interest in 5 years @ 4°/₀?

Prin.

Int. £4 : £210 }
Yrs. 5 :  1  } : : £100 : $x = \dfrac{100 \times 210}{4 \times 5} = £1050.$

For *Years* :  $P = \dfrac{100 \times I}{R \times Y}$   For *Days* :  $P = \dfrac{73000 \times I}{2R \times D}$

52. What principal will produce £384 of Interest in 6 years @ 4°/₀?

53. What principal will produce £153 of Interest in 4½ years @ 4½ °/₀?

54. Find the principal of which the Interest for 50 days @ 4 °/₀ is £14„12.

(10) What Principal will amount to £1260 in 5 years @ 4 °/₀?

Int. on £100 for 5 yr. @ 4°/₀ . . . = £20
Amount of £100 . . . . . . . = £120

Amount.        Prin.
£120 : £1260 : : £100 : $x = \dfrac{100 \times 1260}{120} = £1050.$

For *Years* :  $P = \dfrac{100 \times A}{100 + (R \times Y)}$   For *Days* :  $P = \dfrac{100 \times A}{100 + \frac{R \times D}{365}}$

**63.**  55. Find the principal which in 4½ yrs. @ 4½ °/₀ will amount to £962.

56. What principal will amount to £1017″15 in 6¼ yrs. @ 3 °/₀?

57. What principal lent from March 10 to May 22 @ 5 °/₀ will amount to £712″9″5?

(11) At what rate must £730 be lent for 95 days to amount to £739″10?

Prin. £730 : £100 ⎫  Int.
Da.      95 : 365 ⎭ :: £9″10 : $x = \frac{36500 \times 9\frac{1}{2}}{730 \times 95} = £5.$

For *Years:* $R = \frac{100 \times I}{P \times Y}$    For *Days:* $R = \frac{73000 \times I}{2 \times P \times D}$

58. At what rate must £424 be lent for 2½ yrs. to produce £26″10 of Interest?

59. At what rate must £255″10 be lent from April 1 to June 20, to produce £2″16 of Interest?

(12) Lent £1825 @ 3 °/₀, when will £10″13 of Int. be due?

Prin. £1825 : £100 ⎫  Da.
Int.   £3   : £10″13 ⎭ :: $365 : x = \frac{36500 \times 213}{1825 \times 60} = 71$ Da.

For *Years:* $Y = \frac{100 \times I}{P \times R}$    For *Days:* $D = \frac{73000 \times I}{2 \times P \times R}$

60. How long must £670 be lent to produce £134 @ 5 °/₀.

61. How long must £91″5 be lent to produce £2 of Int. @ 5 °/₀?

62. Lent £511 on Jan. 1, 1856, @ 4¾ °/₀, when will it amount to £517″13?

---

**64.**

# DISCOUNT.

DISCOUNT is a per-centage charged for the payment of money before it is due.

I.

£200.                          *London, March* 15, 1858.

*Three months after date, I promise to pay to Mr William Jones, or order, Two hundred pounds for value received.*                          *James Brown.*

II.

£200.                          *London, March* 15, 1858.

*Three months after date pay to me or order, Two hundred pounds for value received.*

*To Mr James Brown.*                          *William Jones.*

*James Brown.*

**64.**   No. I. is the form of a *Promissory Note*, in which Mr James Brown promises to pay £200 in 3 months after the given date.

No. II. is the form of an *Inland Bill, drawn* by Mr William Jones and sent to Mr James Brown, who on *accepting* it writes his name across the bill, and becomes bound to pay £200 in 3 months after the given date.

If Mr Jones who *holds* the note or the bill cashes it at the bank before it is due, as on April 19, the bank charges discount for advancing the money.

The value of a bill when it is discounted is termed its *Present Value*. The value of a bill when it becomes due is termed its *Future Value*.

We may compare the Present Value and the Future Value to *Ready Money* and *Credit Price*. Goods which may be had on credit for a certain sum may be bought for less ready money. The Credit Price is the Present Value of the goods increased by Interest; the Ready Money is the Future Value diminished by Discount.

The *Bank or Common Discount* is the Interest on the Future Value of the bill.

The *True Discount* is the Interest on the Present Value of the bill.

The Present Value lent out when the bill is discounted amounts to the Future Value when the bill becomes due. The True Discount is the difference between the Future Value and the Present Value.

Common Discount (C. D.) = Int. on Future Value (F. V.)
True Discount (T. D.)   = Int. on Present Value (P. V.)
Hence, C. D. — T. D. = Int. on (F. V. — P. V.)
But, F. V. — P. V. = T. D.
Hence, C. D. — T. D. = Int. on T. D.

The difference between the Common and the True Discount on a bill is = the Interest on the True Discount.

In Great Britain and Ireland, *Three Days of Grace* are given on all bills except those drawn " *at sight*," which are payable on presentation. When a bill, running for a number of months, and dated on the 31st of a month, becomes due in a month having fewer than 31 days, it is nominally due on the *last day* of the month, and legally due on the *third* of next month.

Find the Common and the True Discount on a bill for £200, drawn March 15, 1858, at 3 months; discounted April 19, @ 4 %.

Nominally due, June 15 | From April 19
Legally due, June 18 | to June 18 = 60 days.

I.

Amount or Future Value . . . . . . £200
Common Discount, or Int. on £200 }  . . 1 ″ 6 ″ 3½ $\frac{34}{73}$
for 60 days @ 4 % . . . . }
Net Proceeds . . . . . . . . . £198 ″ 13 ″ 8½ $\frac{39}{73}$

**64.**
II.

$$\text{Int. on £100 for 60 days @ 4 °/}_o = \frac{£100 \times 60 \times 8}{73000} = £\tfrac{4\,4}{7\,3}$$

Future Value.        Present Value.
£100$\tfrac{44}{73}$ : £200 : : £100 : $x$ = £198$_{\prime\prime}$13$_{\prime}$10$\tfrac{1}{4}\tfrac{1434}{1837}$

If we wish the answer correct within a farthing, we may express the fraction decimally, and use contracted division.
£100·6575 : £200 : : £100 : $x$ = £198·693 = £198$_{\prime\prime}$13$_{\prime\prime}$10$\tfrac{1}{2}$.

Amount or Future Value . . . . . . £200
True Net Proceeds or Present Value . .  198$_{\prime\prime}$13$_{\prime\prime}$10$\tfrac{1}{4}\tfrac{1434}{1837}$
True Discount . . . . . . . . . . £ $\overline{\ \ 1_{\prime}\ \ 6_{\prime}\ \ 1\tfrac{1}{4}\tfrac{403}{1837}}$

*Proof.*

True Discount . . . . . . . . . £1$_{\prime\prime}$6$_{\prime\prime}$1$\tfrac{1}{4}\tfrac{403}{1837}$
Int. on the True Disc. for 60 da. @ 4 °/$_o$  0$_{\prime\prime}$0$_{\prime\prime}$2$\tfrac{2}{9}\tfrac{33112}{1537101}$
Common Discount . . . . . . . . £1$_{\prime\prime}$6$_{\prime\prime}$3$\tfrac{1}{4}\tfrac{34}{73}$

Find the *Common* and the *True* Discount on the following bills :

|     |         | Drawn. |             |     |   | Discounted. |     |   |
|-----|---------|--------|-------------|-----|---|-------------|-----|---|
| 1.  | £300    | . . .  | Mar. 25 for 3 months. | . . . | April 16 | @ 4 °/$_o$ |
| 2.  | 600     | . . .  | June 23 | $_{\prime\prime}$ 3 | $_{\prime\prime}$ | . . . | July 15 | $_{\prime\prime}$ 4 | $_{\prime\prime}$ |
| 3.  | 275     | . . .  | Aug. 4 | $_{\prime\prime}$ 2 | $_{\prime\prime}$ | . . . | Aug. 31 | $_{\prime\prime}$ 5 | $_{\prime\prime}$ |
| 4.  | 360     | . . .  | Mar. 19 | $_{\prime\prime}$ 2 | $_{\prime\prime}$ | . . . | April 10 | $_{\prime\prime}$ 3 | $_{\prime\prime}$ |
| 5.  | 275     | . . .  | Mar. 11 | $_{\prime\prime}$ 3 | $_{\prime\prime}$ | . . . | April 1 | $_{\prime\prime}$ 5 | $_{\prime\prime}$ |
| 6.  | 720     | . . .  | Oct. 19 | $_{\prime\prime}$ 2 | $_{\prime\prime}$ | . . . | Nov. 10 | $_{\prime\prime}$ 3 | $_{\prime\prime}$ |
| 7.  | 137$_{\prime}$10 | . . | Mar. 7 | $_{\prime\prime}$ 2 | $_{\prime\prime}$ | . . . | April 3 | $_{\prime}$ 5 | $_{\prime\prime}$ |
| 8.  | 315$_{\prime\prime}$10 | . . | July 10 | $_{\prime}$ 4 | $_{\prime\prime}$ | . . . | Sept. 11 | $_{\prime\prime}$ 3$\tfrac{1}{2}$ | $_{\prime\prime}$ |
| 9.  | 480$_{\prime\prime}$12$_{\prime\prime}$6 | Jan. 1 | $_{\prime\prime}$ 6 | $_{\prime\prime}$ | . . . | Mar. 31 | $_{\prime\prime}$ 4 | $_{\prime\prime}$ |
| 10. | 157$_{\prime\prime}$15 | . . | Nov. 30 | $_{\prime\prime}$ 3 | $_{\prime\prime}$ | . . . | Dec. 30 | $_{\prime}$ 3$\tfrac{1}{2}$ | $_{\prime\prime}$ |
| 11. | 68$_{\prime}$15 | . . | Oct. 31 | $_{\prime\prime}$ 4 | $_{\prime\prime}$ | . . . | Jan. 25 | $_{\prime\prime}$ 5 | $_{\prime}$ |
| 12. | 240$_{\prime\prime}$6$_{\prime}$3 | . | Oct. 31 | $_{\prime\prime}$ 4 | $_{\prime\prime}$ | . . . | Nov. 28 | $_{\prime\prime}$ 4 | $_{\prime\prime}$ |

13. What sum will at the rate of 5 °/$_o$ amount in a year to £75?

14. Find the present worth of £89 due in a year @ 5 °/$_o$.

15. The price of goods, allowing 6 months' credit @ 5 °/$_o$, is £4$_{\prime\prime}$8$_{\prime\prime}$10. Find the ready-money price.

16. What ready money is equivalent to 30/6 with 4 months' credit at 5 °/$_o$?

17. The credit price of a newspaper per annum is £2$_{\prime\prime}$4. Find the ready money payable in advance, taking true discount @ 10 °/$_o$.

18. What sum due in one day will produce 1d. of true discount at 5 °/$_o$?

19. What sum due in one day will produce 1/ of common discount at 5 °/$_o$?

20. Find the *common* discount on a sum for 1 yr. @ 5 °/$_o$, of which the *true* discount for the same time and rate is 5/5.

## 65. EQUATION OF PAYMENTS.

EQUATION OF PAYMENTS shows when a number of debts payable at different times may be adequately paid at once.

(1) Find the equated time for paying £90 due in 80 days, £30 in 92 days, and £120 in 105 days.

$$
\begin{array}{rcl}
£90 \times 80 &=& 7200 \\
30 \times 92 &=& 2760 \\
120 \times 105 &=& 12600 \\
\hline
240 & & )22560(94 \text{ days.} \\
& & 216 \\
& & \overline{\phantom{00}96} \\
& & 96
\end{array}
$$

Suppose 94 days to be the equated time for the payment of the sums mentioned in (1), at the equated time interest will be chargeable on £90 for 14 days, and on £30 for 2 days. But if £120 which is paid 11 days before due be lent out at the same rate, the interest produced by £120 in 11 days would balance the interest chargeable on £90 and £30. 11 days must, however, elapse before this interest can be had, so that the *True Discount* and not the *Interest* on £120 should be = the interest chargeable on £90 and £30 at the equated time.

This approximate method, which is, however, sufficiently accurate for practical purposes, furnishes the correct answer to the following :—

Lent £90 for 80 days, £30 for 92 days, £120 for 105 days. In what time will their sum produce the same interest?

Int. on £90 for 80 days = Int. on £1 for 7200 days.
" 30 " 92 " = " " 2760 "
" 120 " 105 " = " " 12600 "
Total interest . . . = " " 22560 "
= Int. on £240 for 94 days.

(2) £80 is payable *to-day*, £80 in 30 days, £90 in 40 days, £50 in 60 days. Find the equated time.

$$
\begin{array}{rcl}
80 \times 0 &=& 0 \\
80 \times 30 &=& 2400 \\
90 \times 40 &=& 3600 \\
50 \times 60 &=& 3000 \\
\hline
300 & & )9000(30 \text{ days.} \\
& & 9000
\end{array}
$$

**65.** Exercises like (1) may also be performed somewhat similarly, thus :—

$$
\begin{array}{rl}
90 \times 0 = & 0 \\
30 \times 12 = & 360 \\
120 \times 25 = & 3000 \\
\hline
240 \quad & )\overline{3360} \\
& 3360
\end{array}
$$

Days.     Days.

$(14 + 80 = 94$

Find the equated time approximately for paying the following
sums due in the following number of days :—

1. £40 in 54 days, £80 in 36 days.
2. £30 in 58 days, £90 in 26 days.
3. £19 in 12 days, £22 in 24 days, £31 in 36 days.
4. £360 in 15 days, £140 in 20 days, £400 in 17 days.
5. ⅓ of a debt in 6 mo., ₁⁄₄ in 7 mo., ¼ in 8 mo., and the remainder in 9 mo.
6. ⅓ of a debt in 3 mo., ⅓ in 4 mo., and the remainder in 4½ mo.
7. £190 payable *to-day*, £220 in 12 days, £310 in 24 days.
8. £95 payable 3 days *ago*, £110 in 9 days, £155 in 21 days.
9. ⅓ of a debt payable *to-day*, ⅓ in 48 days, and the remainder
in 64 days.

Find the date on which the sum of the following debts can be
adequately paid :—

10. £115 due on Mar. 2 ; £300 on Mar. 20 ; £600 on Mar. 21 ;
£500 on Mar. 29.
11. £30 due on Apr. 1 ; £50 on Apr. 16 ; £30 on Apr. 26 ; £25
on May 1 ; and £15 on May 21.
12. £64 due on Apr. 1 ; £60 on Apr. 13 ; £50 on Apr. 18 ; £30
on Apr. 20 ; £28 on Apr. 24.

---

**66.**             **STOCKS.**

STOCK is the money or capital belonging to any company.
*Government Stocks* consist of the various loans granted to government which form the *National Debt*. The different kinds
of government stock are designated according to the annual
rates of interest they yield ; thus the *Three per cents* yield £3
on every £100 of stock. The price of stock is estimated ⅌
£100 ; thus when the 3¼ ⅌ cents are at 95, the value of £100
stock is £95 sterling.

(1) Find the annual income derived from £450 of stock
in the 3¼ per cents.

**66.**

Stock. Income.

£100 : £450 : : £3¼ : $x$ = £14″12″6.

1. Find the annual revenue derived from £56525 stock in the 3 per cents.

2. Find the annual income obtained from £10,871″10 stock in the 3 per cents.

(2) Find the value of £1350 in the 3 per cents @ 82.

Stock. Sterling.

£100 : £1350 : : £82 : $x$ = £1107.

A person on buying or selling stock per a stockbroker pays ⅛ % of brokerage on the amount of stock.

(3) Find the buying price of £650 stock @ 80¾.

Stock. Sterling.

£100 : £650 : : £(80¾ + ⅛) : $x$ = £525″13″9.

(4) Find the selling price of £825 stock @ 91⅛.

Stock. Sterling.

£100 : £825 : : £(91⅛ — ⅛) : $x$ = £750″15.

3. Find the value of £800 stock @ 95⅛.

4. Find the value of £450 stock @ 88.

5. Find the buying price of £375 stock in the 3 per cents @ 70⅜, allowing brokerage @ ⅛ %.

6. What was paid for £650 stock in the 3½ per cents @ 91¼, allowing brokerage @ ⅛ %?

7. Find the selling price of £330 stock in the 3 per cents @ 87⅛, paying brokerage @ ⅛ %.

8. How much was obtained for £570 stock in the 3½ per cents @ 94⅝, allowing brokerage @ ⅛ %?

(5) Find the quantity of stock @ 92 equivalent to £828.

Sterling. Stock.

£92 : £828 : : £100 : $x$ = £900.

(6) How much stock may be bought for £361 @ 90⅛?

Sterling. Stock.

£(90⅛ + ⅛) : £361 : : £100 : $x$ = £400.

(7) How much stock of the 3 per cents @ 93¾ has realized £1235″17?

Sterling. Stock.

£(93¾ — ⅛) : £1235″17 : : £100 : $x$ = £1320.

66. 9. Find the quantity of stock @ $81\frac{7}{8}$ worth £655.

10. Find the quantity of stock @ $83\frac{3}{4}$ worth £502„10.

11. How much stock @ $93\frac{3}{8}$ may be bought for £750, allowing brokerage @ $\frac{1}{8}$ °/₀?

12. How much stock @ $81\frac{1}{4}$ may be bought for £434, allowing brokerage @ $\frac{1}{8}$ °/₀?

13. Find the quantity of stock @ $96\frac{1}{8}$ which will realize £576, allowing brokerage @ $\frac{1}{8}$ °/₀?

14. Find the quantity of stock @ $92\frac{1}{2}$ which will realize £739, allowing brokerage @ $\frac{1}{8}$ °/₀.

(8) Find the rate of interest obtained from capital invested in the 3 per cents @ $92\frac{5}{8}$.

$$£92\tfrac{5}{8} : £100 :: £3 : x = 3\tfrac{177}{741}\ °/_0.$$

15. Find the rate of interest obtained when the $3\frac{1}{2}$ per cents are @ $95\frac{1}{2}$.

16. What rate of interest is obtained when the $3\frac{3}{4}$ per cents are @ $97\frac{1}{8}$?

(9) How do the $3\frac{1}{4}$ per cents stand when they yield 4 °/₀?

$$£4 : £3\tfrac{1}{4} :: £100 : x = £81\tfrac{1}{4}.$$

17. How do the 3 per cents stand when they yield 4 °/₀?

18. How do the $3\frac{1}{4}$ per cents stand when they yield $3\frac{1}{2}$ °/₀?

(10) Find the annual income derived from a capital of £617„10 invested in the 3 per cents @ 95.

$$£95 : £617„10 :: £3 : x = £19„10.$$

19. What income is derived from a capital of £611„5 invested in the $3\frac{1}{2}$ per cents @ $81\frac{1}{2}$?

20. Find the income derived from £308 invested in the $3\frac{1}{4}$ per cents @ 82.

(11) What sum must be invested in the $3\frac{1}{2}$ per cents @ 85 to produce £24„10 of annual income?

$$£3\tfrac{1}{2} : £24\tfrac{1}{2} :: £85 : x = £595.$$

21. What sum must be invested in the $3\frac{3}{4}$ per cents @ $84\frac{1}{4}$ to produce an income of £50?

22. How much must be invested in the $3\frac{1}{2}$ per cents @ $92\frac{1}{8}$ to produce an income of £504?

23. A legacy of £2000, reduced by a duty of 3 °/₀, has been invested in the $3\frac{1}{4}$ per cents @ $97\frac{3}{8}$. Find the amount of the annually derived income.

24. Bought £300 stock @ $90\frac{1}{2}$, and sold it @ $95\frac{1}{4}$; what was

66. gained, allowing ⅛ % for brokerage on both the buying and the selling price?

25. When the 3 per cents are @ 89, at what rate must the 3½ per cents stand to produce the same rate of interest?

26. Find the difference in the rate of interest between the 3 per cents @ 90 and the 3½ per cents @ 98.

27. A person buys £800 stock @ 91, and sells out @ 93⅛. What does he gain, allowing ⅛ % for brokerage on the buying and the selling price?

28. Invested £1380 in stock @ 91⅞, and sold out @ 90⅛. How much was lost, reckoning the usual brokerage on the buying and the selling price?

---

## 67. PROFIT AND LOSS.

IN PROFIT AND LOSS we consider the difference between the Buying and the Selling prices of commodities.

The *Buying Price* or *Prime Cost* (P. C.) is the sum at which goods are bought; the *Selling Price* (S. P.) is that at which they are sold.

The difference between the buying and selling prices is termed *Gain* or *Loss*, according as the Selling Price is *greater* or *less* than the Prime Cost.

(1) How much is gained by selling 234 yards of cloth @ 6/5½, bought @ 4/3½ ℘ yd.?

| s. d. | | |
|---|---|---|
| 6 ״ 5½ | S. P. ℘ yd. | 234 yd. @ 2/2 |
| 4 ״ 3½ | P. C. ״ ״ | = £25 ״ 7 Total Gain. |
| 2 ״ 2 | G. ״ ״ | |

(2) How much is lost by selling 12 cwt. 3 qr. 16 lb. sugar @ 4½d. ℘ lb., bought @ £2 ״ 4 ״ 4 ℘ cwt.?

| | | |
|---|---|---|
| 4½d. ℘ lb. = £2 ״ 2 | S.P. ℘ cwt. | 12 cwt. 3 qr. 16 lb. @ 2/4 |
| 2 ״ 4 ״ 4 | P.C. ״ ״ | ℘ cwt. = £1 ״ 10 ״ 1 |
| 2 ״ 4 | Loss ״ ״ | Total Loss. |

1. What is gained by selling 367 yards of cloth @ 7/9, bought @ 6/5 ℘ yard?

2. How much is gained by selling 3 cwt. 1 qr. of cheese @ 6½d. ℘ lb., bought at the rate of £2 ״ 6 ״ 8 ℘ cwt.?

3. How much is gained ℘ cwt. by selling sugar @ 5½d. ℘ lb., bought @ £2 ״ 4 ״ 4 ℘ cwt.?

**67** 4. Find the loss on 364 qr. of wheat, bought @ 65/6 ᵽ qr., and sold @ 7/11½ ᵽ bushel.

5. What is gained by selling 10 dozen of pears at two for 1½d., bought at the rate of 5 a-penny?

6. What did a publisher gain by buying the remainder of an edition consisting of 420 copies for £57ⁿ10ⁿ6, and selling 300 copies @ 3/6, and the remaining number @ 3/?

7. Bought 3 cwt. 1 qr. 9 lb. of soap @ £2ⁿ11ⁿ4 ᵽ cwt., and sold it @ 6d. ᵽ lb., but found that the soap had inlaked 27 lb. What was gained or lost by the transaction?

8. Bought 2 cwt. 27 lb. sugar @ 58/4 ᵽ cwt., and sold 1 cwt. 3 qr. @ 7½d. ᵽ lb., but by a fall of the market was obliged to sell the remainder @ 5d. ᵽ lb. What was gained or lost by the transaction?

(3) Find the selling price of 14 cwt. 3 qr. 21 lb. of coffee bought @ £6ⁿ10ⁿ8 ᵽ cwt., and sold with a profit of 5d. ᵽ lb

£6ⁿ10ⁿ8 P. C. ᵽ cwt.    14 cwt. 3 qr. 21 lb. @ £8ⁿ
5d.ᵽ lb.= 2ⁿ 6ⁿ8 G.   ⁿ   ⁿ        17ⁿ4 ᵽ cwt. = £132ⁿ
————————
8ⁿ17ⁿ4 S. P. ⁿ   ⁿ        8ⁿ11 Total S. P.

(4) What must a corn merchant pay for 500 stones of hay, so as to sell it @ 8¼d. with a gain of 1¼d. ᵽ stone?

8¼d. S. P. ᵽ stone.    500 stones @ 7d.
1¼d. G.   ⁿ   ⁿ        = £14ⁿ11ⁿ8 Total P.C.
————
7d. P. C. ⁿ   ⁿ

9. How must 288 yd. of cloth, bought @ 4/5½ ᵽ yd., be sold ᵽ yd. to gain 12 guineas by the transaction?

10. How must 3 pieces of cloth, each 89 yd., bought for £73ⁿ 8ⁿ6, be sold ᵽ yd., to gain £2ⁿ4ⁿ6 ᵽ piece?

11. Find the prime cost of 6 chests of tea, each containing 2 qr. 27 lb., sold @ 4/8 ᵽ lb. with a total gain of £15ⁿ2ⁿ6.

12. At what rate ᵽ cwt. must a merchant purchase a lot of Cumberland hams, so as to retail them @ 9d. ᵽ lb. with a gain of 1¼d. ᵽ lb.?

13. What was paid for 4 cwt. 3 qr. 16 lb. of Cheshire cheese, sold at 6¾d. ᵽ lb. with a gain of 4/8 ᵽ cwt.?

14. At what rate must soap be retailed ᵽ lb. so as to gain 1¼d. ᵽ lb. on 3 cwt. 2 qr. 14 lb., purchased in all for £8ⁿ9ⁿ2?

15. What is the prime cost ᵽ cwt. of 6 cwt. 3 qr. 17 lb. of coffee, sold @ 1/8 ᵽ lb. with a total gain of £12ⁿ1ⁿ6¾?

16. How much does a retailer receive for 3 cwt. 2 qr. of raisins, bought at 42/ ᵽ cwt., and sold with a profit of 2¼d. ᵽ lb.?

**67.** (5) Find the gain % by selling Dutch butter @ 10¼d. ⅌ lb., bought at the rate of 84/ ⅌ cwt.

$$84/ ⅌ \text{ cwt.} = \frac{10\frac{1}{4}\text{d. S. P. } ⅌ \text{ lb.}}{\overline{1\frac{1}{4}\text{d. Gain.}}}$$

9d. : 1¼d. : : 100 : $x$ = 16⅔ %.

(6) Find the loss % by selling 50 copies of a work @ 7/6, 80 copies @ 4/, and the remainder of the edition for £12, the cost of publication being £72″10.

50 copies @ 7/6 = £18″15
80   ″   @ 4/  =  16
Remainder   =  12    £72″10 : £25″15 : : 100 :
S. P . . . . . £46″15    $x$ = 35½⅘ Loss %.
P. C. . . . . 72″10
Loss . . . . £25″15

17. What was gained % by purchasing goods for £16″12″6, and selling them for £17″10″1½?

18. Find the gain % by selling butter @ 7½d. ⅌ lb., bought @ £2″6″8 ⅌ cwt.

19. Bought 2 cwt. 1 qr. 7 lb. soap for £4″17″1½, and sold it @ 5¼d. ⅌ lb. What was gained %?

20. What was lost % on tea, bought @ 2/7 ⅌ lb., duty 2/1 ⅌ lb., and sold @ 4/4 ⅌ lb.?

21. Bought 37 yd. of cloth @ 13/6 ⅌ yd., sold 34½ yd. @ 16/, and the remnant @ 2/6 below prime cost. What was gained %?

22. Bought 3 cwt. 3 qr. of coffee for £23″12″6, but on account of damage was obliged to sell one-half @ 1/1 ⅌ lb., and the other half @ 1/ ⅌ lb. What was lost %?

23. How much does a photographer gain % by buying frames @ 29/6 ⅌ doz., and selling them @ 4/6 each?

24. Bought a sloop for £180, paid £40 for new mast and anchor, sold her for £275. What was gained %, allowing ½ % on the selling price for commission agency?

25. Bought 26 cwt. 2 qr. 14 lb. of cheese @ 52/ ⅌ cwt.; sold 20 cwt. wholesale @ £3″10 ⅌ cwt., and retailed the remainder @ 9d. ⅌ lb. What was gained %?

26. A picture-seller who paid £250 for engraving a picture, sold 12 India proofs @ 3 guineas each, and 240 prints @ £1″11″6 each. What was gained % by the transaction?

(7) At what rate must cheese, bought @ 50/ ⅌ cwt., be sold ⅌ lb. so as to gain 12 %?

**67.** $100 : 112 :: 50/ : x = 56/$ S. P. ℔ cwt.
$= 6d.$ S. P. ℔ lb.

(8) Find the buying price of cloth, sold @ 9/6 ℔ yd. with a loss of 24 %?

$$\begin{array}{r} 100 \\ \underline{24} \\ 76 \end{array} : 100 :: 9/6 : x = 12/6 \text{ P. C. ℔ yd.}$$

27. At what rate must starch bought @ 42/ ℔ cwt. be sold ℔ lb. so as to gain 33⅓ %?

28. Find the prime cost of coffee ℔ cwt. sold @ 1/10 ℔ lb. with a profit of 10 %.

29. What was the prime cost of goods sold for £26„5 with a loss of 12½ %?

30. Bought 7 cwt. 3 qr. Java rice for £4„10„5. How must it be sold ℔ lb. to gain 20 %?

31. Find the prime cost of a work of 10 vols, sold @ 10/6 ℔ vol. with a profit of 16⅔ %.

32. A contractor gains 16⅔ % by performing a piece of work for £233„19„5. What is his outlay for workmanship and materials?

33. A paper merchant bought 100 reams of foolscap, and sold 50 reams @ £1„5, with a gain of 11⅑ %; 25 reams @ £1„8; and the remainder, being damaged, @ 17/8. Find the total prime cost, and the gain or loss %.

34. Find the weekly outlay of the proprietor of an omnibus who receives on an average £3„15„3 every lawful day, and thus clears 75 %.

35. At what price must cloth bought @ 5/6 ℔ yd. be rated so as to allow 4 % discount for ready money and gain $9\frac{1}{11}$ % by the money received?

36. Suppose a bootmaker pays on an average 6/4 for the leather and furnishings of a pair of boots, and 6/4 for the workmanship; what must he charge his customer so as to allow him a discount of 5 %, and gain 50 % by the money received?

(9) Sold goods for £225„10 with a gain of 12¾ %. What would have been gained or lost % by selling them for £187„10?

$$£225„10 : £187„10 :: 112\tfrac{3}{4} : x = \begin{array}{r} 100 \\ \underline{93\tfrac{3}{4}} \\ 6\tfrac{1}{4} \end{array} \text{ Loss \%}$$

(10) Sold a bale of leather for £14„14, and gained 17⅗ %. How should it have been sold to have gained 18 %?

$117\tfrac{3}{5} : 118 :: £14„14 : x = £14„15$ S. P.

**67.** 37. A bookseller having bought two copies of the seventh edition of the Encyclopædia Britannica at the same price, sold one @ £25 with a profit of 9$\frac{1}{7}$ %. How much did he gain % by selling the other @ £27*10?

38. A merchant of Lyons by selling silk @ 10 francs $\gamma$ metre gained 20 %. What did he lose % by selling silk of the same prime cost @ 8 francs $\gamma$ metre?

39. Lost 36 % by selling cloves @ 8d. $\gamma$ lb. What would have been gained or lost % by selling them @ 1$\frac{1}{4}$d. $\gamma$ oz.?

40. Gained 13 % by selling paper @ 9/5 $\gamma$ ream. What was lost % by selling paper of the same value @ 8/3 $\gamma$ ream?

41. Sold a bale of leather for £15, and lost 25 %. How should it have been sold to have gained 33 %?

42. Sold pencils at the rate of 3 for 2d., and gained 33$\frac{1}{3}$ %. What would have been gained or lost % by selling them @ 5$\frac{1}{4}$d. $\gamma$ doz.?

43. A bootmaker by selling boots @ 24/ $\gamma$ pair gains 50 %. What must he have charged to have given a discount of 5 %, and to have gained 78$\frac{1}{2}$ %?

(11). Find the prime cost and selling price of goods sold with a gain of 32 %, and of £16*17*4 in all.

$$32 : 100 : : £16\text{\it *}17\text{\it *}4 : x = £52\text{\it *}14\text{\it *}2 \text{ P. C.}$$
$$\underline{16\text{\it *}17\text{\it *}4 \text{ Gain.}}$$
$$£69\text{\it *}11\text{\it *}6 \text{ S. P.}$$

44. Sold goods with a loss of 20 %, and lost £57*6*8 by the transaction. What was the prime cost?

45. Find the selling price of goods by which there was a loss of 2 % or of £54*10 by the whole transaction.

46. What does a draper receive for 39 yd. of cloth which he sells with a gain of 2/ $\gamma$ yd. and of 26$\frac{2}{3}$ %?

47. Sold cheese with a gain of 2$\frac{1}{2}$d. $\gamma$ lb. or of 62$\frac{1}{2}$ %. At what was it bought and sold $\gamma$ cwt.?

48. Sold 39 casks of cod-liver oil, each containing 52$\frac{1}{2}$ gallons, with a loss of 1$\frac{3}{4}$ %, and of £8*10*7$\frac{1}{2}$ on the transaction. What was the prime cost $\gamma$ gallon?

49. Find the original outlay of a publisher who sold 2000 copies of a guide-book, with a gain of 6d. $\gamma$ copy and of 25 %.

50. Find the outlay of a publisher who sells 500 prints of an engraving with a gain of 5/6 $\gamma$ print and of 35$\frac{11}{13}$ %.

(12) How much sugar, bought @ £2*13*8 $\gamma$ cwt. was sold @ 5d. $\gamma$ lb., with a total loss of £3*18*9?

**67.**

$$£2\text{"}13\text{"}8 \not{v} \text{ cwt.} = 5\tfrac{3}{4}\text{d. P. C. } \not{v} \text{ lb.} \qquad £3\text{"}18\text{"}9$$
$$5 \text{ d. S. P.} \qquad \text{"} \qquad \overline{78\text{s.}}$$
$$\overline{\tfrac{3}{4}\text{d. Loss.}} \quad \text{"} \qquad \overline{945\text{d.}}$$
$$3)3780 \qquad\qquad 3780\text{f.}$$
$$\overline{1260} \text{ lb.} = 11 \text{ cwt. 1 qr.}$$

(13) How many prints of an engraving must a picture-dealer sell @ £1"11"6, so that he may gain $51\tfrac{1}{5}$ % on an outlay of £250?

$$100 : 151\tfrac{1}{5} : : £250 : x = £378 \text{ S. P.}$$
$$£1\text{"}11\text{"}6 \qquad\qquad £378$$
$$\overline{31} \text{ s.} \qquad\qquad \overline{7560} \text{ s.}$$
$$\overline{63} \text{ sixd.} \qquad )\overline{15120} \text{ sixd. (240 prints.}$$

51. Bought a cargo of oranges @ 12/6 ꝗ chest, and sold it with a gain of 30 %, and of £18"15 in all. How many chests were in the cargo?

52. How many yd. of cloth bought @ 13/2¼ ꝗ yd. must a draper sell @ 16/6 to gain £3"19"6?

53. What quantity of butter bought @ £2"13"8 ꝗ cwt. must be sold @ 7½d. ꝗ lb. to clear £4"18?

54. Bought haddocks @ 3/4 ꝗ long hundred (120). How many must be sold at 7d. ꝗ dozen to gain 12/6?

55. How much sugar bought @ 42/ ꝗ cwt. must be sold @ 6d. ꝗ lb. to gain £20 in all?

56. Bought 10 cwt. of sugar @ 44/ ꝗ cwt., and sold it at 4½d. ꝗ lb.. How much tea bought @ 3/1 ꝗ lb. must be sold @ 4/4 ꝗ lb. to cover the loss on the sugar?

57. Sold iron @ £5"6 ꝗ ton, with a profit of 6 %, and of £21"10"6 in all. What quantity was sold?

58. A drysalter purchases goods @ 58/4 ꝗ cwt., and by retailing them gains £2"17"6½, being at the rate of 4 %. What quantity was sold?

59. A grocer buys sugar @ 37/4 ꝗ cwt., and by selling it @ 62½ % profit gains £5"5"5. What quantity does he sell?

60. Bought a cargo of oranges @ 15/ ꝗ chest, and sold one-half of them @ 19/6 ꝗ chest, and the other with a loss of 10 %, but gained £27"7"6 on the whole. How many chests were bought?

(14) Bought goods for £53, and sold them for £75, with one year's credit. What was gained %?

Let us first find the *Present Value* of £75, reckoning the rate of interest here and in all the following examples at *Five per cent.*

**67.**   £105 $=$ Future Value of £100 in 1 yr. @ 5 $°/_{°}$.
  105 : 100 : : £75 : $x = £71\frac{3}{7}$ P. V. of S. P.
  The question is now reduced to the following :—*Bought goods for £53, and sold them for £71$\frac{3}{7}$ $°/_{°}$; what was gained $°/_{°}$?*
    £53 : £71$\frac{3}{7}$ : : 100 : $x = 134\frac{2\,4\,0}{3\,7\,1}$ S. P.
      Gain $°/_{°} = 34\frac{2\,4\,0}{3\,7\,1}$.
  These two statements may be united as follows:
$$\left. \begin{array}{l} 105 : 100 \\ £53 : £75 \end{array} \right\} : : 100 : x$$

(15) How must cloth, bought @ 6/9 $\Psi$ yd., with 3 months' credit, be sold so as to gain 5 $°/_{°}$, and allow 9 months' credit.

F. V. of £100 @ 5 $°/_{°}$ for 3 mo. and 9 mo. $= £101\frac{1}{4}$ and £103$\frac{3}{4}$.
$$\left. \begin{array}{l} 101\frac{1}{4} : 103\frac{3}{4} \\ 100 \ : 105 \end{array} \right\} : : \begin{array}{c} \text{s. d.} \\ 6\text{\textit{''}}9 \end{array} : x = \begin{array}{c} \text{s. d.} \\ 7\text{\textit{''}}3\frac{3}{2\,0} \end{array} \text{ S. P.}$$

61. Bought goods for £59, and sold them for £89 with one year's credit; what was gained $°/_{°}$?

62. What was lost by selling 288 yards of cloth for £182″8, bought 6 months ago @ 12/6 $\Psi$ yd.?

63. Bought goods for £70, and sold them for 70 guineas with twelve months' credit; what was gained or lost $°/_{°}$?

64. How must goods be sold to gain 5 $°/_{°}$, and give 9 months' credit, bought the same day for £81 with 3 months' credit?

65. What is gained or lost $°/_{°}$ by selling goods @ £47″13″4 $\Psi$ cwt. bought 6 months ago @ 8/ $\Psi$ lb.?

66. What is lost $°/_{°}$ by selling goods with 6 months' credit, bought 6 months ago for the same money?

**68.**   DISTRIBUTIVE PROPORTION.

In Distributive Proportion we divide or *distribute* a given number into parts which have a given ratio to each other.

(1) Divide £376″5 of gain among three partners in an adventure whose risks are respectively £225, £150, and £250.

  £225
  150
  250
  ———
  625 : 225 : : £376″5 : $x$ $=$  £135″ 9
  625 : 150 : :  376″5 : $x$ $=$    90″ 6
  625 : 250 : :  376″5 : $x$ $=$   150″10
                                  ————
                                  £376″ 5

**68.** The sum of the risks = £625. *As* the whole risk *is to* each risk, *so is* the sum to be divided *to* the share of each. The sum is thus divided into *parts proportional* to 225, 150, and 250, which may be cancelled by their common factor 25.

The following method is often convenient :—

| 225 | 9 |
|---|---|
| 150 | 6 |
| 250 | 10 |

$$9 \times £15\prime\prime1 = 135\prime\prime\ 9$$
$$6 \times \ 15\prime\prime1 = \ 90\prime\prime\ 6$$
$$10 \times \ 15\prime\prime1 = 150\prime\prime10$$

25)£376″5(£15″1       £376″ 5

(2) A sum of £1000 was bequeathed to four relations, and by an inadvertency in the will, it was stated that they were to receive $\frac{1}{2}$, $\frac{1}{3}$, $\frac{1}{4}$, and $\frac{1}{6}$ of the sum respectively. How much should each receive according to the spirit of the will?

| | | |
|---|---|---|
| $\frac{1}{2} = \frac{6}{12}$ | 6 | $6 \times £66\prime\prime13\prime\prime4 = £400$ |
| $\frac{1}{3} = \frac{4}{12}$ | 4 | $4 \times \ 66\prime\prime13\prime\prime4 = \ 266\prime\prime13\prime\prime4$ |
| $\frac{1}{4} = \frac{3}{12}$ | 3 | $3 \times \ 66\prime\prime13\prime\prime4 = \ 200$ |
| $\frac{1}{6} = \frac{2}{12}$ | 2 | $2 \times \ 66\prime\prime13\prime\prime4 = \ 133\prime\ 6\prime\prime8$ |

15)£1000(£66″13″4      £1000

We divide £1000 in the mutual ratios of $\frac{1}{2}$, $\frac{1}{3}$, $\frac{1}{4}$, $\frac{1}{6}$. The sum of these fractions = $\frac{15}{12}$ is greater than unity. $\frac{1}{12}$ is therefore *one-fifteenth* of the sum. Dividing £1000 by 15, we multiply by 6, 4, 3, 2, successively to obtain the respective shares.

1. Divide 84 into parts having the mutual ratios of 2, 3, 7.

2. Divide 1200 into parts having the mutual ratios of 11, 12, 13, 14.

3. Divide a line 4 feet long into parts having the ratios of the first four odd numbers.

4. Divide 100 into parts having the ratios of the cubes of the first three numbers.

5. Divide 390 into parts having the ratios of $\frac{1}{2}$, $\frac{1}{3}$, $\frac{1}{4}$.

6. Divide 1321 into parts having the ratios of the *reciprocals* of the first three even numbers.

7. Apportion a house tax of £6″18″8 among 3 joint proprietors, who pay in the proportion of the annual values of their properties, which are £30, £40, and £60 respectively.

8. A vessel is divided into 64 equal shares, of which A, B, C, D, have 6 shares each; E, 12; F, 16; G, 4; and H the remainder. Find their respective shares in sustaining a joint loss of £158″10″1.

9. Divide a profit of £689 among 3 partners, of whom the first owns $\frac{2}{13}$ of the joint stock and the second $\frac{5}{13}$.

10. A, B, C, D, invest £450, £230, £190, and £110 respectively

**68.** in a speculation. Find their respective liabilities in a joint loss of £313″12.

11. Three partners respectively claim $\frac{1}{3}$, $\frac{1}{15}$, and $\frac{7}{18}$ of the gain of an adventure amounting to £1260. Give to each a proportionate share.

12. Divide 5 guineas among George, James, and Henry, who respectively claim $\frac{3}{4}$, $\frac{1}{5}$, and $\frac{1}{6}$, so that they may have proportionate shares.

13. An analysis of the manure of dissolved bones gives the following results for every 100 parts: — Water, 13·97; Organic Matter, 15·71; Soluble Phosphates, 21·63; Insoluble Phosphates, 11·43; Sulphate of Lime, 15·83; Sulphuric Acid, 15·63; Alkaline Salts, 1·10; Silica, &c., the remainder. Find the weight of each in a ton of dissolved bones.

14. Oil of vitriol (HO, $SO_3$) contains by weight, 1 of Hydrogen, 32 of Oxygen, and 32 of Sulphur. Find the weight of each in a gallon of oil of vitriol which weighs 18½ lb.

(3) D, E, and F, gain £564: D's capital of £300 has been in trade for 6 months; E's, which is £400, for 3 mo.; F's, which is £500, for 2 mo. Find the share of each.

D, £300×6=1800,9          9×£28″4=£253″16
E,  400×3=1200'6          6×£28″4=  169″ 4
F,  500×2=1000|5          5×£28″4=  141
                £
        20)564                    £564
        £28″4

The use of £300 in trade for 6 mo. is equivalent to that of 6 times £300 for 1 mo. Similarly, £400 for 3 mo. is equivalent to 3 times £400 for 1 mo.; and £500 for 2 mo. to 2 times £500 for 1 mo. Taking the time of 1 month alike for D, E, F, we see that the shares are proportional to 1800, 1200, and 1000.

(4) A commences trade with £3000: in 3 months B joins him with £4000; at the end of the next 2 months A takes out £1000; in 1 mo. after C joins them with £2000, and B adds £1500; in 2 mo. after C takes out £500: at the end of 12 months they divide £2760 of gain. What is the share of each?

A has £3000 in trade for 5 mo., and £2000 for 7 mo.
B ″  £4000      ″    ″  3 ″   and £5500 ″ 6 ″
C ″  £2000      ″    ″  2 ″   and £1500 ″ 4 ″

$$A \begin{cases} £3000 \times 5 = 15000 \\ 2000 \times 7 = 14000 \end{cases} 29{,}000$$

$$B \begin{cases} £4000 \times 3 = 12000 \\ 5500 \times 6 = 33000 \end{cases} 45{,}000$$

**68.**

$$C \left\{ \begin{array}{l} £2000 \times 2 = 4000 \\ 1500 \times 4 = 6000 \end{array} \right\} 10,000$$

29 × £32″17″1½ ⁴⁄₇ =£ 952″17″1½ ⁴⁄₇
45 × 32″17″1½ ⁴⁄₇ = 1478″11″5⁹⁄₇ ⁴⁄₇
10 × 32″17″1½ ⁴⁄₇ = 328″11″5⁹⁄₇ ⁴⁄₇

84)£2760                    £2760″ 0″0
————
£32″17″1½ ⁴⁄₇

15. In a copartnery, A's capital of £400 has continued for 9 mo.; B's of £350 for 8 mo.; C's of £600 for 2 mo. Divide £570 of gain among them.

16. Three cattle-dealers rent a field of 9 acres @ £5 ℔ acre: A puts in 6 cows for 2 months; B, 9 cows for 1 mo.; C, 12 cows for 3 mo. How much does each pay?

17. At the end of 12 months, D, E, F, having a joint capital of £6000, find that they have lost £625. D's capital of £2500 has been in trade for 12 mo., E's of £1500 for 8 mo., and F's for 4 mo. What is the loss of each?

18. A and B enter into partnership, the former with £1800, the latter with £900: in 8 months B adds £300 to his capital. Divide a profit of £840 between them at the end of 12 months.

19. A has £300 in trade for 7 months, when B joins him with £400. At the end of the next 3 months C joins them with £300. Divide £549 of gain among them after 18 months' trade.

20. A, B, and C, enter into partnership on Jan. 1, 1856, with a capital of £1000 *each*. On April 30, B withdraws £400, and C makes up the sum. On Aug. 28, A withdraws £200, and C makes up the sum. On balancing their books for the year they find they have a gain of £365. What is the share of each?

21. Three graziers rent a field from May 11 to October 19, 1857, for £43. A agrees to pay £13 for grazing 12 oxen; B, £18 for 18 oxen; and C the remainder for 20 oxen. To how many days is each grazier entitled; and if the oxen go into the field in the order A, B, C, on what days do B's and C's severally enter?

☞ The times are proportional to the sums paid for 1 ox. A pays £¹³⁄₁₂; B, £¹⁸⁄₁₈; C, £¹²⁄₂₀ for 1 ox.

22. 3 men and 4 boys are loading carts with sand. A man takes 7 shovelfuls for a boy's 6, and 4 shovelfuls of a man's = 5 of a boy's. Divide £3″7 proportionally among them.

## 69.        ALLIGATION.

ALLIGATION treats of the prices and quantities of a compound and its ingredients.

In *Alligation Medial*, the prices of the ingredients are given, and the price of the compound is obtained by finding the *average* price.

(1) A merchant mixes 45 gallons of spirits @ 7/4, 20 @ 6/6, 84 @ 6/8, and 21 gallons of water. What is the price of the compound ℔ gal.?

$$\begin{array}{rcl}
45 \text{ @ } 7/4 &=& 330\text{s.} \\
20 \text{ .. } 6/6 &=& 130 \\
84 \text{ .. } 6/8 &=& 560 \\
\underline{21} & & \\
170 & & )\,\overline{1020}\,(\,6/ \text{ ℔ gal.}
\end{array}$$

The average is thus found by multiplying each price by the corresponding quantity, and finding the sum of the products by the sum of the quantities.

1. Find the average price of 4 gal. @ 5/, 5 @ 4/, 8 @ 2/6, and 7 @ 3/.

☞ We may thus often find the average price merely, without considering that the whole has been compounded.

2. Find the average price of 100 lb. rice @ 1d. ℔ lb., 300 lb. @ 2d., 400 @ 1½d., and 100 @ 4d.

3. Find the price ℔ gal. of a mixture of spirits of 50 gal. @ 4/6, 40 @ 4/2, 45 @ 4/4.

4. Find the average price of 23 qr. wheat @ 40/, 32 @ 48/, 12 @ 69/, 24 @ 38/, and 17 @ 50/.

5. On Feb. 6, 1856, the following quantities of wheat were sold at the six highest prices in the Edinburgh Grain Market:—8 quarters @ 96/; 4 @ 84/; 21 @ 78/; 13 @ 76/; 1 @ 75/; 2 @ 74/. Find the average price ℔ qr. as deduced from these prices and quantities.

In *Alligation Alternate*, we find the proportional quantities of ingredients of given prices which will produce a compound of a given price.

(2) Mix spirits @ 8/3, 7/9, 6/6, and 8/4 ℔ gal. so that the compound may be worth 8/ ℔ gal.

**I.**

$$96 \begin{cases} 78 \\ 93 \\ 99 \\ 100 \end{cases}$$

| d. | gal. | | | |
|---|---|---|---|---|
| 78 | 3 | × 78 | = | 234 |
| 93 | 4 | × 93 | = | 372 |
| 99 | 18 | × 99 | = | 1782 |
| 100 | 3 | × 100 | = | 300 |
| | 28 | | 28) | 2688 |
| | | | | 96 |

**II.**

| d. | gal. | | | |
|---|---|---|---|---|
| 78 | 4 | × 78 | = | 312 |
| 93 | 3 | × 93 | = | 279 |
| 99 | 3 | × 99 | = | 297 |
| 100 | 18 | × 100 | = | 1800 |
| | 28 | | 28) | 2688 |
| | | | | 96 |

We express the prices in the *same* name.

To obtain a compound at 96d. we must mix two ingredients, of which the one is dearer and the other cheaper than the compound.

We may, as in Method I., connect 78d. with 99d., and 93d. with 100d.

The act of thus *connecting* or *binding* the prices *together* is the reason why the rule is termed *Alligation*.

If spirits worth 99d. ℔ gal. are sold @ 96d. there will be a loss of 3d., and if spirits worth 78d. are sold @ 96d. there will be a gain of 18d. Since 18 × 3d. = 3 × 18d., the loss on 18 gallons worth 99d. will balance the gain on 3 gallons worth 78d. We therefore write the difference between 96 and 78 or 18 opposite its *alternate* number 99; and the difference between 99 and 96 or 3 opposite its *alternate* number 78. We proceed similarly with 93d. and 100d.

In Method II. we may connect 99d. with 93d. and 78d. with 100d.

When the differences between the price of the compound and that of a dearer and of a cheaper ingredient connected together are *equal*, we may take *any equal quantity* of each of the latter; thus, instead of 3, 3, 4, 18, we may take $x$, $x$, 4, 18, where $x$ may be any quantity.

6. Find the proportional quantities of sugar @ 5d. and 8d. that must be sold to make the average price 7d. ℔ lb.

7. What proportional quantities of potatoes @ 2/, 3/, and 3/6 ℔ bushel must be sold to make the average price 2/9 ℔ bushel?

8. Mix tea @ 4/6, 4/2, 3/4, and 3/9 ℔ lb., so that the compound may be worth 3/11 ℔ lb.

9. What proportional quantities of wine @ 15/, 12/, 18/, 19/, and 21/ ℔ gal. must be sold to make the average price 16/ ℔ gal.?

(3) What quantities of tea @ 5/3, 4/5, and 2/9, must be mixed with 21 lb. @ 6/1, to make the whole worth 5/ ℔ lb.?

**69.**

I.

$$60\begin{cases}33 \\ 53 \\ 63 \\ 73\end{cases}\quad \begin{array}{l} 3 \times 3 = 9 \; @ \; 33d. \\ 13 \times 3 = 39 \; .. \; 53d. \\ 27 \times 3 = 81 \; .. \; 63d. \\ 7 \; ..... \; 21 \; .. \; 73d. \end{array}$$

II.

$$60\begin{cases}33 \\ 53 \\ 63 \\ 73\end{cases}\quad \begin{array}{l} 13 + 3 = 16 \times \frac{3}{4} = 9\frac{3}{4} \; @ \; 33d. \\ 13 \quad = 13 \times \frac{21}{34} = 8\frac{1}{34} \; .. \; 53d. \\ 27 \quad = 27 \times \frac{3}{4} = 16\frac{3}{4} \; .. \; 63d. \\ 27 + 7 = 34 \; ..... \; 21 \; .. \; 73d. \end{array}$$

Having found the proportional quantities as formerly, we multiply them by the *ratio* of the given quantity to its corresponding proportional quantity.

Similarly, when the quantity of the compound is given, we multiply the proportional quantities by the ratio of the given quantity to the *sum* of the proportional quantities.

10. How much wheat @ 42/ and 56/ must be sold with 13 qr. of wheat @ 60/ to make the average price 50/ ℞ qr.?

11. How much sugar @ 10d. and 11d. must be mixed with 9 lb. of 7d. sugar to make the whole worth 8¼d.?

12. How many gallons of water must be mixed with 63 gallons of spirits @ 8/ so that the prime cost may be 7/ ℞ gal.?

☞ We *alligate* 8/ with 0. Or we may solve this by proportion,

s.   s.   gal.   gal.

7 : 8 : : 63 : 72. ∴ Number of gal. of water = 72 — 63.

13. How many gallons of water must be mixed with 47¼ gallons of spirits @ 6/3 to make the prime cost 5/ ℞ gal.?

14. How many gallons of each kind of wine @ 15/3, 16/4, 17/2, and 18/1, must be sold to make the average price of 154 gallons 17/ ℞ gal.?

15. The Specific Gravity of an alloy of gold and copper is 16·65, while that of gold is 19·2, and that of copper 9. Find the weight of gold and copper in 144 oz. of the alloy.

16. A crown made of gold and silver weighs 150 oz. and displaces 13·824 cub. in. of water. Had it been gold it would have displaced 12·96 cub. in. of water, and had it been silver it would have displaced 23·04 cub. in. Find the weight of gold and silver in the crown.

☞ This question is founded on the story of Archimedes and Hiero. Hiero had given a goldsmith a certain quantity of gold to make a crown. In course of time, the artificer presented a crown of the same weight as that of the quantity of gold; but as Hiero suspected a fraud, he requested Archimedes to discover if any baser metal had been alloyed with the gold. Archimedes considered that if the crown contained any metal lighter than gold, it would be larger than a pure gold crown of the same weight. Having obtained a mass of pure gold and of the other metal, each of the same weight as the crown, he found the quantity of water which each of the three displaced, and from these data discovered the proportion of each metal in the crown.

## 70.       BARTER.

In Barter, two parties mutually give goods of *equal* value in exchange.

(1) Exchanged 164 lb. of tea @ 4/8 ℔ lb. for coffee @ 1/7 ℔ lb. How many lb. of coffee were received?

$x$ lb. of coffee @ 1/7 = 164 lb. of tea @ 4/8.

$$x = \frac{164 \times 56}{19} = 483\tfrac{7}{19}\ \text{lb.}$$

(2) In return for 146 qr. wheat @ 70/ ℔ qr., an agent received Wilts cheese @ 88/ ℔ cwt., and Dunlop cheese @ 60/ ℔ cwt., obtaining 6 cwt. of Wilts for every 5 of Dunlop. How many cwt. of each were received?

$$6 \times \overset{s.}{88} = \overset{s.}{528}$$
$$5 \times 60 = 300$$

828s. = the price of 1 parcel of both kinds of cheese in the given proportional quantities.

$x$ parcels @ 828s. = 146 qr. @ 70/ = 10220s.

$$x = \frac{146 \times 70}{828} = \tfrac{10220}{828} = 12\tfrac{71}{207}\ \text{parcels.}$$

Each parcel contains 6 cwt. of Wilts and 5 cwt. of Dunlop.

cwt. qr. lb.
$$6 \times 12\tfrac{71}{207} = 74\tfrac{4}{69}\ \text{cwt.} = 74\text{·}0\text{·}6\tfrac{34}{69}\ \text{Wilts.}$$
$$5 \times 12\tfrac{71}{207} = 61\tfrac{148}{207}\ \text{cwt.} = 61\text{·}2\text{·}24\tfrac{16}{207}\ \text{Dunlop.}$$

*Proof* $\begin{cases} 74\tfrac{4}{69}\ \text{cwt.} \ @\ 88/ = 6517\tfrac{7}{69}\text{s.} \\ 61\tfrac{148}{207}\ \text{cwt.} \ @\ 60/ = 3702\tfrac{62}{69}\text{s.} \end{cases}$
$$\overline{10220\text{s.}}$$

1. How many yd. of cloth @ 2/3 are worth 54 lb. of tea @ 4/1?

2. What is the price ℔ yd. of cloth, of which 200 yd. are worth 2 cwt. 2 qr. 25 lb. @ 93/4 ℔ cwt.?

3. How many gallons of brandy @ 24/6 ℔ gal. are worth 35 doz. loaves of refined sugar, each 16 lb. @ 70/ ℔ cwt.?

4. Exchanged a tierce of sugar weighing 8 cwt. 3 qr. 14 lb. for 31 cwt. 0 qr. 7 lb. rice @ 18/ ℔ cwt. Find the price of the sugar ℔ lb.

5. How many yd. of linen cambric @ 5/6 must be given in exchange for 15 dozen pairs of boots @ 18/ ℔ pair, and 13 dozen pairs of shoes @ 8/ ℔ pair?

**70.** 6. A baker, who has run an account with a grocer for 12¼ lb. tea @ 4/2, 60 lb. sugar @ 6½d., 3¼ lb. coffee @ 1/8, and 13 drums of sultana raisins, each 20 lb., @ 11d. ℣ lb., has a contra-account of 23 dozen loaves @ 7½d. ℣ loaf. How many loaves @ 8½d. will settle the account?

7. A dairyman, who has supplied a baker with 90 pints of milk @ 2½d., 13½ pints of cream @ 10d., and 80 lb. of butter @ 10d., agrees to take an equal number of loaves @ 7d. and 7½d. How many of each does he get?

8. Exchanged 28 lb. of tea @ 4/2 for coffee, and got 5 lb. of coffee for 2 lb. of tea. How many lb. of coffee were got, and what was its price ℣ lb.?

9. In return for 80 qr. barley @ 56/ ℣ qr., ⅜ of the value was received in bone-dust @ £8„8 ℣ ton, and the rest in money. How much money and how many tons of bone-dust were received?

10. In return for 165 cwt. flour @ 15/ ℣ cwt., an agent received 3 chests of tea, each 81 lb., @ 4/4 ℣ lb., and 8 doz. loaves of refined sugar, each 19¾ lb. What was sugar ℣ lb.?

11. In return for 14 cwt. 2 qr. 20 lb. Glo'ster cheese @ 77/ ℣ cwt.; beef @ 8d. ℣ lb., and mutton @ 7d. ℣ lb., were received in the ratio of 7 lb. of beef for every 3 lb. of mutton. How much of each was received?

12. Exchanged 6 cwt. 2 qr. 3 lb. salmon @ 1/6 ℣ lb., 20 turbots @ 4/2, 16 dozen haddocks @ 4/6 ℣ doz., and 15 pints of shrimps @ 6d., for 2 cows @ £9„13 each, 160 lb. beef @ 7½d., 240 lb. pork @ 5d., and 80 pairs of fowls @ 3/9 ℣ pair. How many lb. of mutton @ 7d. must be given for the balance?

---

**71.**

# CHAIN RULE.

(1) IF 5 pheasants are worth 4 grouse; 5 grouse, 8 partridges; 2 partridges, 5 snipes; how many snipes may be had for 10 pheasants?

$x$ snipes     = 10 pheasants
5 pheasants =   4 grouse
5 grouse     =   8 partridges
2 partridges =   5 snipes

Having arranged the *pairs of equal values* or *equations*, so that numbers of the same name are on different sides, we examine the equations as follows:—

G

**71.**

| | Snipes. | | | Snipes. |
|---|---|---|---|---|

1 partridge $= \frac{5}{2}$     $\left.\begin{array}{l}\text{4 grouse}\\\text{or 5 pheasants}\end{array}\right\} = \frac{5\times 8\times 4}{2\times 5}$

$\left.\begin{array}{l}\text{8 partridges}\\\text{or 5 grouse}\end{array}\right\} = \frac{5\times 8}{2}$     1 pheasant $= \frac{5\times 8\times 4}{2\times 5\times 5}$

1 grouse $= \frac{5\times 8}{2\times 5}$     10 pheasants $= \frac{5\times 8\times 4\times 10}{2\times 5\times 5}$

We see then that the number of snipes $=$ 10 pheasants is obtained by dividing the product of the numbers on one side by the product of those on the other.

N$^{\text{o}}$ of snipes $= \frac{5\times 8\times 4\times 10}{2\times 5\times 5}$, which by cancelling $=$ 32.

This method is known as the CHAIN RULE. Each equation is a *link* in the chain; each link begins with the *name* with which the preceding link ended, and the chain is complete when the last ends with the name in the first link, whose number is wanted.

(2) How many francs are $=$ a lac of 100,000 rupees, each 1/10¼; 25·22 francs being $=$ £1.

$$x \text{ francs} = 100{,}000 \text{ rupees}$$
$$1 \text{ rupee} = 89 \text{ f.}$$
$$960 \text{ f.} = 25\cdot 22 \text{ francs}$$

$$x = \frac{252200\times 89}{96} = 233810\tfrac{5}{12} \text{ francs.}$$

1. 9 old ale gallons $=$ 11 old wine gallons of which 9 $=$ 20 Scotch pints, and 8 Scotch pints $=$ 3 Imperial gallons. How many Imperial gallons $=$ 54 ale gallons?

2. How many Linlithgow barley firlots $=$ 3 Winchester bushels of which 33 $=$ 32 Imperial bushels or Linlithgow wheat firlots, and 16 Linlithgow wheat firlots $=$ 11 Linlithgow barley firlots?

3. How many Scotch acres $=$ 100 Irish acres, 121 Irish acres $=$ 196 Imperial acres, and 126 Imperial acres $=$ 100 Scotch acres?

4. 8 Scotch miles $=$ 9 Imperial miles; 14 Imperial miles $=$ 11 Irish miles. How many Irish miles $=$ 112 Scotch miles?

    ☞ The mutual ratios in the preceding examples are convenient *approximations.*

5. 2 quarts of plums are worth 3 of pears; 6 of pears $=$ 5 of apples; 8 of apples cost 2/4. Find the price of 3 quarts of plums.

6. In 1855, the mutual ratios of the weights of bales of cotton imported at Liverpool from the following places were as follow:— 2 from Bombay $=$ 3 from Egypt; 9 Brazil $=$ 4 United States; 7 Brazil $=$ 5 Egypt; 7 Calcutta $=$ 5 Madras; 14 United States $=$ 15 Madras. How many from Calcutta were $=$ 50 from Bombay?

**71.** 7. By examining the average weight of the bales of cotton imported at Liverpool in 1843, the following were obtained:—55 from Egypt = 69 from W. Indies; 35 from Alabama = 43 from the Upland U. States, from which 207 = 350 from Egypt; 91 from Alabama = 215 from Brazil, from which 27 = 13 from E. Indies. How many from W. Indies = 165 from E. Indies?

8. From the Imperial averages for the week ending 30th April 1857, it appeared that the price of 39 quarters of barley = that of 73 of oats; 68 of barley = 73 of beans; 27 of beans = 28 of pease; 39 of wheat = 58 of rye, of which 153 = 143 of pease. How many quarters of wheat = 638 of oats?

9. 4 talents were = 375 lb. avoir., and each talent contained 3000 shekels. Find the weight of a shekel in oz. avoir.

10. 273 quarters of wheat = 638 of oats, of which 73 = 39 of barley, sold @ 42/7 ⅌ quarter. Find the price of 1 quarter of wheat.

11. By a comparison of the apothecaries' grains of different countries, it was found that 17 German = 20 British; 85 German = 86 Neapolitan; 37 Spanish = 45 Austrian; and 185 Spanish = 172 Neapolitan. How many British = 90 Austrian?

12. A mile = 80 chains = 63360 inches; a chain = 100 links. How many inches are in a link?

13. 175 lb. troy = 144 lb. avoir., each 7000 grains, of which 3608 = 1 Cologne mark. How many Cologne marks = 451 lb. troy?

14. If a metre = 39·37079 in. be taken as $\frac{1}{40,000,000}$ of the earth's circumference, how many miles are in the earth's circumference?

15. 4 nautical miles = a German mile; the earth's circumference contains 5400 German miles = 40,000,000 metres. How many feet are in a nautical mile?

**72.** # EXCHANGE.

EXCHANGE is the method of changing the money of one country into that of another.

The *Par of Exchange* is the *real* comparative value of the money of two countries, estimated by the weight and fineness of the coins.

The *Course of Exchange* is the comparative value of the money of two countries, which fluctuates according to the circumstances of commerce.

In Exchange, £1 is generally adopted as the unit of comparison. Thus, the par of exchange with France is 25 francs 22½ centimes ⅌ £1. When £1 is the unit, the equivalent in *foreign* money *varies*

**72.**  in the course of exchange; thus, £1 may be exchanged at one time for 24 fr. 30 c., and at another for 25 fr. 50 c. When a foreign coin is taken as the unit, the equivalent in *sterling varies* in the course of exchange; thus, while the par with Naples is 39¾d. ৺ ducat, the exchange may at one time be 38d., and at another 40d. ৺ ducat.

CANADA.—Accounts are kept in £, s. D. *Currency*, of which £1, being taken as = 4 dollars of the *Nominal* value of 4/6 each, is = 18/ sterling. Hence the nominal par is £100 currency = £90 sterling. But as the real average value of the dollar is 4/2, £1 currency = 16/8 sterling, and the real par is £108 currency = £90 sterling. The *Nominal Par* is taken as the standard, and a *Premium* is added to show the course of exchange. At a premium of 8 %, £108 currency=£90 sterling.

WEST INDIES.—The old currencies are now superseded by sterling. Of the foreign coins in circulation, the principal are the dollar = 4/2, and the doubloon = £3„4.

(1) How much sterling is = £327 currency, at a premium of 9 %?

*By the Chain Rule.*

Currency.      Sterling.

£109 : £327 :: £90 : $x$     or $\begin{cases} \text{Ster.} & x = £327\,\text{Curr.} \\ \text{Curr.} £109 = £90 \text{ Ster.} \end{cases}$

$$x = \frac{327 \times 90}{109} = £300 \text{ Ster.}$$

(2) How much currency is = £8460 sterling, at a premium of 9½ %?

Sterling.          Currency.

£90 : £8460 :: £109½ : $x$   or $\begin{cases} \text{Curr.} & x = £8460\,\text{Ster.} \\ \text{Ster.} £90 = £109½\,\text{Curr.} \end{cases}$

$$x = \frac{8460 \times 109½}{90} = £10293 \text{ Curr.}$$

1. How much currency will an emigrant to Canada receive for £135„7„6 sterling, at a prem. of 8½ %?

2. An emigrant on arriving at Toronto changes 6 crowns, 7 hf.-crowns, 37 shillings, and 5 sixpences sterling, to currency, @ the rate of 15d. for 1/ sterling. How much currency does he receive?

3. How much sterling is = £324„2„3 currency, remitted from Montreal, at a premium of 8 %?

4. An agent at Quebec wishes to remit to his employer in London £489„12„1½d. To how much sterling will this be equal at a premium of 8½ %?

5. How many dollars @ 4/2, or how many doubloons @ £3„4, must a Jamaica merchant receive from his correspondent at Cuba, who is due £320?

**72.**

6. An agent changes 3000 dollars to sterling @ 4/2 ℔ dollar, at Kingston, Jamaica, on embarking for Halifax, Nova Scotia, and on arriving changes the sterling to currency @ 8 °/₀ premium. How much currency does he receive?

UNITED STATES.—Accounts are kept in dollars and cents. 10 *cents** = 1 dime; 10 dimes= 1 *dollar** ($); 10 dollars = 1 eagle. The par of exchange, deduced from the gold coins, is $1 = 4/1¼ nearly; from the silver coins, $1 = 4/2¼ nearly. In custom-house valuations, $1 = 4/2. The nominal par of exchange is $1 = 4/6; hence, $40 = £9, or $100 = £22″10. We take the nominal par as the standard, and add a premium to $100; thus, at a premium of 9½ °/₀, $109½ = £22″10.

(3) How much sterling is in $111·55, at a premium of 9⅛ °/₀?

$$109\overset{\$}{\cdot}125 : 11\overset{\$}{1}·55 : : £22″10 : x = £23.$$

(4) How many $ are in £2560, at a premium of 9 °/₀?

$$2\overset{£}{2}″10 : 25\overset{£}{6}0 : : 1\overset{\$}{0}9 : x = \$12401·77\tfrac{7}{9}.$$

7. How much sterling is in $3390, remitted from New York, at a premium of 8 °/₀?

8. How much sterling is in $994·25, remitted from Philadelphia, at a premium of 9½ °/₀?

9. How many $ are received at Boston for £738, at a premium of 9⅞ °/₀?

10. How many $ are in £7659, received at New Orleans, at a premium of 10 °/₀?

The following Illustrative Processes may suffice for the rest of the Exercises:—

(5) Change £999″12 to florins at Vienna, at the rate of 10 florins 50 kreuzers ℔ £1.

$$£1 : £999″12 : : 1\overset{\text{fl.}}{0}″5\overset{\text{kr.}}{0} : x = 10829\,\text{fl. or} \begin{cases} \text{fl. } x = £999·6 \\ £1 = 650\,\text{kr.} \\ \text{kr. } 60 = 1\,\text{fl.} \end{cases}$$

(6) A merchant remits 717 thalers 12 groschen from Berlin, at the rate of 6 thalers 24 groschen ℔ £1. To how much sterling is this equivalent?

$$6\overset{\text{th.}}{″}2\overset{\text{gr.}}{4} : 71\overset{\text{th.}}{7}″1\overset{\text{gr.}}{2} : : \overset{£}{1} : x = £105″10 \text{ or} \begin{cases} £ x = 717·4\,\text{th.} \\ \text{th. } 6·8 = £1 \end{cases}$$

---

* The names of the coins quoted in Exchanges are put in *Italics*.

**72.**    (7) How many dollars may be had at Malaga for £809″15″10, at 4/2 ℔ dollar?

$$4/2 : £809″15″10 :: 1 : x = 3887 \quad \text{or} \begin{cases} \text{doll}.x = 194350\,\text{d.} \\ \text{d. } 50 = 1 \text{ doll.} \end{cases}$$

(8) What is the value in sterling of 733 oncie, remitted from Palermo, @ 10/5 ℔ oncia?

$$1 : 733 :: 10.5 :: £381″15″5 \quad \text{or} \begin{cases} £x &= 733 \text{ onc.} \\ \text{onc.} 1 &= 125 \text{ d.} \\ \text{d.} 240 &= 1 \text{ £} \end{cases}$$

FRANCE; BELGIUM.—Accounts are kept in francs and centimes. 100 *centimes* = 1 *franc* = 9¼d. nearly. The par of exchange, deduced from the gold coins, is 25 fr. 22¼ c. per £1; and from the silver coins, 25 fr. 57 c. per £1. The franc weighs 5 grammes, and is coined of silver $\frac{9}{10}$ fine.

11. How many francs are = £525″10″6, remitted to Marseilles, @ 25 fr. 22 c. ℔ £1?

12. How many francs must be remitted from Brussels to pay a bill of £987″14″6, @ 25 fr. 10 c. ℔ £1?

13. How much sterling must be remitted to Paris to settle an account of 9900 francs, @ 24 fr. 75 c. ℔ £1?

14. How much sterling must be sent to Antwerp to be equivalent to 25663 fr. 75 c., @ 24 fr. 50 c. ℔ £1?

HOLLAND.—5 *cents* = 1 stiver; 20 stivers or 100 cents = 1 *florin* or *guilder* = 1/8. Par of exchange, 12 florins = £1.

15. How many florins must be paid at Amsterdam in order to liquidate a debt of £1500″8; exch. 12 fl. 6 c. ℔ £1?

16. A merchant at Gouda consigns cheese to the amount of 8993 florins to an agent in Scotland. How much sterling must the latter remit @ 11 fl. 50 c. ℔ £1?

SWITZERLAND.—10 rappen = 1 batz; 10 batzen = 1 *franc* = 1/2 nearly. The French coinage is also used.

17. A London jeweller remits £701″12″6 to a watchmaker in Geneva @ 25 fr. 30 c. ℔ £1. How many francs does the latter receive?

18. A merchant of Geneva, on coming to Berne, changes 518 French to Swiss francs @ 148 French for 100 Swiss francs. How many Swiss francs does he receive?

AUSTRIA.—4 pfennings = 1 *kreuzer;* 20 kreuzers = 1 zwanziger; 3 zwanzigers or 60 kreuzers = 1 *florin* = 2/0¼ nearly. Par of exchange, 9 fl. 50 kr. = £1.

**72.** 19. How many florins will be received at Vienna for £786″14″6, exch. 10 fl. 30 kr. ♯ £1?

20. The sum of 19868 fl. is remitted *from* Augsburg in Bavaria, where the Austrian coinage is used. Find the value in sterling @ 10 fl. 45 kr. ♯ £1.

SOUTHERN GERMANY.—4 pfennings=1 *kreuzer* ; 60 kreuzers = 1 *florin* = 1/8 nearly. Par of exchange, 120½ fl. = £10.

21. How many florins will be received at Frankfort-on-the-Maine for £767, exch. 119¼?

22. An agent at Munich remitted 2241 florins. Find the value in sterling, exch. 124¼.

PRUSSIA ; HANOVER, &c.—12 pfennings = 1 *groschen;* 30 groschen = 1 *thaler* = 2/10¾ nearly. Par of exchange, 6 thal. 27 gr. = £1.

23. How many thalers may be had at Dantzic for £726″15″6, exch. 6 th. 20 gr.?

24. How much sterling is = 36473 th. 20 gr. remitted from Hanover, exch. 6 th. 10 gr.?

*Bremen.*—5 schwaren = 1 grote; 72 grotes = 1 *rixdollar* = 3/3½ nearly. Par of exchange, 609 r. d. = £100.

25. How many r. d. may be had at Bremen for £575, exch. 608?

26. How much sterling is = 1517 r. d. 36 gr. remitted from Oldenburg, where the Bremen coinage is used, exch. 607?

*Hamburg ; Lubec.*— 12 pfennings = 1 *schilling ;* 16 schillings = 1 *mark*. Money is distinguished into *Banco* and *Currency*. *Banco* is used in Hamburg in exchanges in wholesale transactions and in *Bank* business. *Currency* consists of coins in circulation ; the marks current of Hamburg and Lubec are, from the latter, termed marks *Lub*. The *agio* or difference between banco and currency varies from 20 to 25 %. Par of exchange, 13 mk. 10½ sch, banco = £1; 1 mark banco = 1/5½ nearly; 1 mark current = 1/2¼ nearly.

27. How many mk. banco are = £876″8, exch. 13 mk. 4 sch.?

28. How much sterling will be received in London for 27783 mk. banco, remitted through the Bank of Hamburg, exch. 13 mk. 12½ sch.?

29. A merchant pays 6461 mk. curr. into the Bank of Hamburg. How much banco is entered on the books, agio being 24¼ %? (124¼ mk. curr. = 100 mk. banco.)

30. A wholesale merchant in Hamburg debits his agent at Lubec to the amount of 2400 mk. banco. To how many mk. Lub. is this equivalent, agio 21⅜ %?

**72.** DENMARK.—16 *skillings* = 1 mark ; 6 marks = 1 *Rigsbank dollar* = 2/2½ nearly. Par of exchange, 9 R. d. 10 sk. = £1.

NORWAY.—24 *skillings* = 1 mark ; 5 marks = 1 *species dollar* = 4/5 nearly. Par of exchange, 4 sp. d. 53 sk. = £1.

SWEDEN.—48 *skillings* = 1 *rixdollar* banco = 1/8 nearly. Exchanges are generally effected through Hamburg.

31. How many Rigsbank dollars are in £432, remitted to Copenhagen ; exch. 9 R. d. 10 sk. ₳ £1 ?

32. How many species dollars are in £1050″10, remitted to the Bank of Norway at Trondheim ; exch. 4 sp. doll. 53 sk. ₳ £1 ?

33. How much sterling is = 5300 species dollars, remitted through a branch of the Bank of Norway at Bergen ; exch. 4 sp. doll. 50 sk. ₳ £1 ?

34. How much sterling is = 740 rixdollars banco, sent from Stockholm ; exch. 12 r. d. 16 sk. banco ₳ £1 ?

RUSSIA.—100 *copecs* = 1 *ruble* = 3/1¼.

35. A British merchant sends £867″14″6 to an agent at St Petersburg ; what does the latter receive @ 3/1¼ ₳ ruble ?

36. How much sterling must be remitted to Riga to discharge a bill of 1200 Ro 50 c. @ 3/1½ ?

PORTUGAL.—1000 reas = 1 *milrea* (𝕭) = 4/9¼ nearly.

37. How much sterling is = a conto or 1000 𝕭, remitted from Oporto @ 56d. ₳ 𝕭1 ?

38. How many 𝕭 are = £2270, sent to Lisbon @ 56¾d. ₳ 𝕭1 ?

SPAIN.—34 maravedis = 1 real vellon ; 20 reals vellon = 1 *hard dollar* = 4/2 nearly.

*Gibraltar.*—16 quartos = 1 real current ; 12 reals current = 1 *hard dollar.*

39. A soldier on landing at Gibraltar changed 23 hf. sov. to dollars @ 50d. ₳ dollar. How many did he receive ?

40. How much sterling must be remitted to Madrid to discharge an account of 1230 reals @ 50¼d. ₳ dollar ?

AUSTRIAN ITALY.—100 *centesimi* = 1 *lira* = 8¼d. nearly. Par of exchange, 29 l. 52 c. = £1 or 48¾d. = 6 Austrian lire.

SARDINIA.—100 *centesimi* = 1 *lira nuova* = 9½d. nearly = French franc. Par of exchange, 25 l. 22 c. = £1.

TUSCANY.—100 *centesimi* = 1 *lira* = 7¾d. nearly. Par of exchange, 30 l. 68 c. = £1.

41. How many Austrian lire are = £375″10, remitted to Milan ; exch. 29 l. 52 c. ₳ £1 ?

**72.** 42. How many Austrian lire are = £341"5, remitted to Venice; exch. 48¾d. for 6 Austrian lire?

43. How much sterling is in 2300 lire nuove, remitted from Genoa; exch. 25 l. 30 c. ⅌ £1?

44. How much sterling is in 4590 Tuscan lire, remitted from Florence; exch. 30 l. 60 c. ⅌ £1?

ROMAN STATES.—10 bajocchi = 1 *paolo*; 10 paoli = 1 *scudo* or crown = 4/2 nearly. Par of exchange, 48 paoli or pauls = £1.

45. On visiting Rome, an Englishman changes £37"12"6. How many pauls does he receive at the rate of 47½ pauls ⅌ £1?

46. How much sterling is = 3697 scudi, 68 baj., remitted from Ancona, at the rate of 48 pauls ⅌ £1?

NAPLES.—100 grani = 1 *ducat* = 3/3¾ nearly.
SICILY.—600 grani = 1 *oncia* = 10/3½ nearly.

47. A merchant in Naples receives a bill from London to the amount of £861. To how many ducats is this equal; exch. 41d. ⅌ ducat?

48. How much sterling is = 846 oncie; exch. 123d. ⅌ oncia?

TURKEY.—40 paras=1 *piastre*=2d.; about 120 piast.=£1.
EGYPT.—40 paras=1 *piastre*=2¾d.; " 100 " =£1.

49. A traveller pays an interpreter at Constantinople the sum of 500 piastres. What is the value in sterling at 120 piastres ⅌ £1?

50. Change £125"10 to piastres at Alexandria @ 97½ piastres ⅌ £1.

GREECE.—100 *lepta* = 1 *drachma* = 8¾d. nearly. Par of exchange, 28 dr. 15 lp. = £1.

51. Find the difference in Sterling and in Greek money between £44"16 and 1317 dr. 42 lp., exch. at par.

EAST INDIES.—12 pice = 1 anna; 16 annas = 1 *rupee* = 1/10¼.

52. A Calcutta merchant makes a payment of a lac or 100,000 rupees. Find the amount in sterling @ 1/10¼.

53. Sent to Bombay goods worth £299"16"3. To how many rupees is this equivalent @ 1/10½ each?

CHINA.—1000 le or cash = 1 *leang* or *tael*, reckoned by the East India Company @ 6/8. 720 taels = 1000 dollars of 4/9½ nearly.

54. How much sterling is = 5400 taels, paid at Canton, reckoning them @ 6/6 each?

55. A merchant of Hongkong sells goods to the amount of £846 "13"4. How many taels does he receive @ 6/8 each?

**72.** INDIRECT EXCHANGES between two countries are effected through the medium of another. It is seldom that the medium is effected through more than one intermediate place.

(9) How much sterling must be paid in London to pay 749 Rigsbank dollars in Copenhagen through the medium of Hamburg; exch. 13 mk. 6 sch. banco=£1; 200 R. d. = 300 mk. banco.

$$\pounds \ x \ = \ 749 \text{ R. d.}$$
$$\text{R. d. } 200 \ = \ 300 \text{ mk. b.}$$
$$\text{mk. b. } 1 \ = \ 16 \text{ sch.}$$
$$\text{sch. } 214 \ = \ \pounds 1$$
$$x = \frac{749 \times 300 \times 16}{214 \times 200} = \pounds 84.$$

56. How many francs = £250, sent to Paris through Hamburg; exch. 13 mk. 14 sch. banco = £1; 185 fr. = 100 mk.?

57. Find the number of mk. curr. = £180, remitted through Hamburg; exch. 13 mk. 12 sch. banco = £1; agio 20 %.

58. How much sterling must be remitted to Berne through Paris to be equivalent to 6325 Swiss francs; exch. 25 fr. 30 c. ꝑ £1; and 148 French = 100 Swiss francs?

59. How much sterling is = 60,180 paras; exch. between Constantinople and Vienna, 210 paras = 1 florin; between Vienna and London, 9 fl. 50 kr. = £1?

60. How much sterling is = 530 th. 7½ gr.; exchange between Berlin and Paris, 3 fr. 60 c. ꝑ 1 th.; between Paris and London, 25 fr. 20 c. ꝑ £1?

## **73.** INVOLUTION.

INVOLUTION is the continued multiplication of a number by itself.

The continued product thus obtained is termed a *Power* of the given number; and the number of times the number is used as a factor denotes the *Index* of the power. Thus $2 \times 2 \times 2 \times 2 \times 2 \times 2 = 64 = $ *sixth* power of $2 = 2^6$.

(1) Find the *seventh* power of 27.

$$27 \times 27 \times 27 \times 27 \times 27 \times 27 \times 27 = 10,460,353,203$$

Instead of multiplying by the number successively, we may use those powers of which the sum of the indices is equal to the index of the required power; thus,

**73.**

$$27 \times 27 \times 27 = \ldots 19{,}683 \ldots = 27^3$$
$$19683 \times 19683 = 387{,}420{,}489 = 27^3 \times 27^3 = 27^6$$
$$387420489 \times 27 = 10{,}460{,}353{,}203 = 27^6 \times 27 = 27^7$$

Find the following powers:—

| | | | |
|---|---|---|---|
| 1. $17^3$ | 4. $98^6$ | 7. $11^7$ | 10. $13^{10}$ |
| 2. $32^6$ | 5. $99^6$ | 8. $15^8$ | 11. $309^6$ |
| 3. $36^6$ | 6. $101^6$ | 9. $14^9$ | 12. $1002^4$ |

(2) Find the *sixth* power of $\frac{3}{11}$.

$$3^6 = 729 \qquad (\tfrac{3}{11})^6 = \tfrac{729}{1771561}.$$
$$11^6 = 1771561$$

| | | | |
|---|---|---|---|
| 13. $(\tfrac{2}{3})^3$ | 15. $(\tfrac{9}{7})^5$ | 17. $(\tfrac{2}{5})^8$ | 19. $(\tfrac{3}{5})^7$ |
| 14. $(\tfrac{9}{10})^4$ | 16. $(\tfrac{1\frac{1}{2}}{})^6$ | 18. $(\tfrac{4}{15})^5$ | 20. $(\tfrac{23}{37})^4$ |

(3) Find the 5th power of $1 \cdot 025$ true to 6 decimal places (see § 39.)

$$1 \cdot 025 \times 1 \cdot 025 = 1 \cdot 050625$$
$$1 \cdot 050625 \times 1 \cdot 025 = 1 \cdot 076891$$
$$1 \cdot 076891 \times 1 \cdot 050625 = 1 \cdot 131408$$

| | | |
|---|---|---|
| 21. $1 \cdot 04^4$ to 4 pl. | 24. $1 \cdot 025^6$ to 6 pl. | 27. $2 \cdot 625^8$ to 6 pl. |
| 22. $1 \cdot 05^6$ .. 6 .. | 25. $1 \cdot 045^7$ .. 7 .. | 28. $3 \cdot 165^6$ .. 4 .. |
| 23. $1 \cdot 03^7$ .. 4 .. | 26. $1 \cdot 035^{10}$.. 7 .. | 29. $9 \cdot 999^4$ .. 4 .. |

30. Find the area of a floor $19\frac{3}{4}$ ft. square.*

31. Find the cubic content of a die whose side is $1\frac{1}{4}$ inch.

32. How many sq. ft. are contained in the *aroura*, 50 Greek ft. (each $1 \cdot 01146$ ft.) square?

33. How many sq. yd. are in the *are*, 10 *metres* (each $39 \cdot 37079$ in.) square?

34. Find how many cub. ft. are in the *stère* or *cubic metre*.

35. How many flagstones 14 in. square will be required to floor a kitchen 21 ft. square?

36. Find how many cubes $1\frac{1}{4}$ inch in the side can be cut out of 7 cub. ft. 74 cub. in., allowing 3 cub. in. for waste.

*Circles are proportional to the* squares *of their diameters.*

37. How many times is a circle 27 ft. in diameter as large as another 15 in. in diameter?

38. The paving of a circular floor $25 \cdot 6$ ft. in diameter cost £9„13„4; what cost the paving of a similar floor $38 \cdot 4$ ft. in diameter?

---

* When we say a surface is $19\frac{3}{4}$ ft. *square*, we mean it contains $19\frac{3}{4}$ × $19\frac{3}{4}$ *square* ft. A surface 10 ft. *square* is 10 times as large as a surface containing 10 *square* ft.

**73.**    *Spheres are proportional to the* CUBES *of their diameters.*

39. The weight of a metallic ball $\frac{1}{2}$ inch in diameter is ·398 oz. Find the weight of another of the same metal $\frac{3}{4}$ inch in diameter.

40. A ball $\frac{3}{4}$ inch in diameter displaces ·128 oz. of water; how many oz. will another $2\frac{1}{4}$ in. in diameter displace?

41. How many times is the Earth, whose mean diameter is 7912 miles, as large as the Moon, whose diameter is 2140 miles?

*A body, in falling, traverses* 16·1 *ft. during the first second,* 4 × 16·1 *ft. in two seconds, and so on, the* SPACES *traversed being proportional to the* SQUARES OF THE TIMES.

42. Through what space will a body fall in $2\frac{1}{4}$ seconds?

43. To what height must an aeronaut ascend so that a ball let fall from his balloon may reach the ground in the quarter of a minute?

I. *The square of the sum of two numbers is = the sum of their squares increased by twice their product.**

$$9+7$$
$$9+7$$
$$\overline{9^2 \;+9{\times}7}$$
$$\phantom{9^2+}9{\times}7\;+7^2$$
$$\overline{9^2+2(9{\times}7)+7^2}$$

Thus, $(9+7)^2 = 9^2 + 2 \times 9 \times 7 + 7^2$
or $\quad 16^2 = 81 + 126 + 49 = 256.$

Similarly, $(40+3)^2 = 40^2 + 2 \times 40 \times 3 + 3^2$
or $\quad\quad 43^2 = 1600 + 240 + 9 = 1849.$

II. *The square of the difference of two numbers is = the sum of their squares diminished by twice their product.**

$$9-7$$
$$9-7$$
$$\overline{9^2 \;-9{\times}7}$$
$$\phantom{9^2}-9{\times}7\;+7^2$$
$$\overline{9^2-2(9{\times}7)+7^2}$$

Thus, $(9-7)^2 = 9^2 - 2 \times 9 \times 7 + 7^2$
or $\quad 2^2 = 81 - 126 + 49 = 4.$

Similarly, $(50-4)^2 = 50^2 - 2 \times 50 \times 4 + 4^2$
or $\quad\quad 46^2 = 2500 - 400 + 16 = 2116.$

III. *The product of the sum and the difference of two numbers is = the difference of their squares.**

$$9+7$$
$$9-7$$
$$\overline{9^2 \;+9 \times 7}$$
$$\phantom{9^2}-9 \times 7 -7^2$$
$$\overline{9^2 \phantom{+++++} -7^2}$$

Thus, $(9+7) \times (9-7) = 9^2 - 7^2$
or $\quad 16 \times 2 = 81 - 49 = 32.$

Similarly, $(50+7) \times (50-7) = 50^2 - 7^2$
or $\quad\quad 57 \times 43 = 2500 - 49 = 2451.$

---

* These propositions are more conveniently remembered in their algebraic form.

$$\text{I. } (a+b)^2 = a^2 + 2ab + b^2$$
$$\text{II. } (a-b)^2 = a^2 - 2ab + b^2$$
$$\text{III. } (a+b)(a-b) = a^2 - b^2$$

**73.** From III., we obtain a convenient method for obtaining the square of a number *mentally.*\*

By III. $(77 + 3) \times (77 - 3)$ $= 77^2 - 3^2$
Hence $(77 + 3) \times (77 - 3) + 3^2$ $= 77^2$
or $80 \times 74 + 3^2$ $= 77^2$

Find the difference between the given number and a number near it ending in 0. Take a third number, so that the difference between it and the given number may be $=$ the former difference. The square of the given number is $=$ the product of the other two numbers increased by the square of the common difference.

Thus, $93^2 = 90 \times 96 + 9 = 8649$.

From I., we obtain a method applicable when the number to be squared ends in 5 or $\frac{1}{2}$.

By I. $75^2 = 70^2 + 2 \times 70 \times 5 + 5^2$
$= 70 \times 70 + 10 \times 70 + 5^2$
$= 80 \times 70 + 25$

When the last figure is 5, the square may be found by multiplying the number of tens by the next greater number, and then affixing 25. Similarly, $(9\frac{1}{2})^2 = 9 \times 10 + \frac{1}{4} = 90\frac{1}{4}$.

Square the following numbers *mentally :—*

| | | | | | |
|---|---|---|---|---|---|
| 1. 21 | 5. 56 | 9. 72 | 13. 85 | 17. 195 | 21. 19$\frac{1}{2}$ |
| 2. 61 | 6. 89 | 10. 97 | 14. 75 | 18. 895 | 22. 22$\frac{1}{2}$ |
| 3. 33 | 7. 74 | 11. 82 | 15. 35 | 19. 395 | 23. 17$\frac{1}{2}$ |
| 4. 47 | 8. 68 | 12. 64 | 16. 65 | 20. 495 | 24. 25$\frac{1}{2}$ |

# EVOLUTION.

EVOLUTION is that process by which we find a number which when multiplied a certain number of times by itself, reproduces a given number. The number found is termed a *root* of the given number.

## SQUARE ROOT.

**74.** The SQUARE ROOT of a number, when multiplied by itself, reproduces the original number; thus, 3 is the *square root* of 9, $3 = \sqrt{9} = 9^{\frac{1}{2}}$; $8 = \sqrt{64} = 64^{\frac{1}{2}}$.

Take any number, as 43, we know that $43^2 = (40 + 3)^2$
$= 40^2 + 2 \times 40 \times 3 + 3^2$.

---

\* This is said to be the method employed by Margaret Cleland of Darvel, Ayrshire, who has acquired some celebrity for her arithmetical powers.

**74.** Let us now in re-    40   | $40^2+2\times40\times3+3^2(40+3$
producing the number          | $40^2$
determine the method   $2\times40+3$ | $\overline{\phantom{xx}}$
of finding the Square          | $\quad 2\times40\times3+3^2$
Root.                        | $\quad 2\times40\times3+3^2$

Subtracting $40^2$, we leave $2 \times 40 \times 3 + 3^2$. Further to
obtain the quotient 3, the divisor must be $2 \times 40 + 3$.

No number containing 1 figure can have more than 2 figures in its square. No number containing 2 figures can have
more than 4 figures in its square. Since 1 place in a number
corresponds to a period of 2 places in its square, before extracting the square root, *we point off in periods of two places,
commencing at units' place.*

(1) Find $\sqrt{1849}$.

The greatest square root in 18 is 4. Sub-    4 $)18,49($ 43
tracting $4^2$, we have 2, which with the next      16
period annexed is 249. Doubling 4, we see   83, $\overline{249}$
that 8 in 24 is 3 times. Annexing 3 to 8, we     249
subtract $3 \times 83$, and having no remainder, find   $\overline{\phantom{xxx}}$
$43 = \sqrt{1849}$.

We have first subtracted $40^2 = 1600$; we have then subtracted $2 \times 40 \times 3 + 3^2$ or $(2 \times 40 + 3) \times 3 = 249$, to
make up $43^2$.

(2) Find $\sqrt{127449}$.

The greatest sq. root in 12 is 3. Sub-   3    $)12,74,49(357$
tracting 9, and annexing the next period,         9
we have 374. Doubling 3, we see that   65   $\overline{374}$
as we have a figure to annex to 6, the         325
next figure in the quotient will be 5.   707   $\overline{4949}$
Subtracting $5 \times 65$, and taking down the         4949
next period, we have 4949. Adding 5        $\overline{\phantom{xxx}}$
to the divisor, we obtain 70, the double of 35. The next figure being 7, we subtract $7 \times 707$ or 4949; and thus find that
$357 = \sqrt{127449}$.

We have first subtracted $300^2$ or 90,000. Having then
subtracted $2 \times 300 \times 50 + 50^2$ or $(2 \times 300 + 50) \times 50$, we
have now subtracted in all $(300 + 50)^2$ or $350^2$. We then
subtract $2 \times 350 \times 7 + 7^2$ or $(2 \times 350 + 7) \times 7$, and thus
complete the square of 357.

In extracting the square root, no remainder can be greater
than twice the root obtained.

Thus, in finding the greatest square root in a number to be 8, it
is evident the number is less than $(8 + 1)^2$ or $8^2 + 2 \times 8 + 1$.
When $8^2$ is subtracted, the remainder is therefore less than $2 \times 8$

**74.**                    Find the square root of:

| 1. | 1024 | 9. | 7365796 | 17. | 80568576 |
|----|------|-----|---------|-----|----------|
| 2. | 4225 | 10. | 27415696 | 18. | 62473216 |
| 3. | 3136 | 11. | 20820969 | 19. | 88887184 |
| 4. | 137641 | 12. | 14235529 | 20. | 22992025 |
| 5. | 50625 | 13. | 16232841 | 21. | 56987401 |
| 6. | 401956 | 14. | 70207641 | 22. | 58415449 |
| 7. | 5499025 | 15. | 31843449 | 23. | 236144689 |
| 8. | 9897316 | 16. | 79263409 | 24. | 998876025 |

(3). Find $\sqrt{672 \cdot 35675}$ to 5 decimals.

```
2      )6,72·35,67,5(25·92984         2      )6,72·35,67,5(25·92984
       4                                     4
45     272                           45     272
       225                                  225
509    4735                          509    4735
       4581                                 4581
5182   15467                         5182   15467
       10364|                               10364
51849  5103|50                       5,1,8,4 5103
       4666|41                               4666
518588 437|0900                              437
       414|8704                              415
5185964 22|219600                            22
        20|743856                            21
         1|475744                             1
```

In extracting the square root of a number, we need only
extract as many figures as the number next greater than half
the number of the required figures. In the example before
us, we require 5 decimals, and as there are 2 integral places
in the root, there will thus be 7 figures in all. We need only
extract 4 figures, and then finish as in Contracted Division
(see § 40.)

Let us now examine the closeness of the approximation. In
comparing the *first* part of the root which is extracted, with the
*second* part which is required, we must attend to *local value*, by
adding as many ciphers to the former as will give it 7 figures, the
required number in the root.

When the square of the first is subtracted, the remainder is =
twice the product of the first and second with the square of the
second. We now merely divide this by twice the first, so that the
quotient = the second with the square of the second divided by
twice the first. Now the second contains 3 places, hence its square
contains no more than 6 places ; and as twice the first cannot con-

**74.** tain less than 7, the square of the second, divided by twice the first, is a proper fraction, and hence less than 1, so that the quotient is a convenient approximation to the second part of the root.*

(4) Find $\sqrt{\cdot 009}$ to 6 places.

```
9  ) ·0090(·094868
      81
184   900
      736
1888  16400
      15104
18,9,6 1296
       1138
       158
       152
       6
```

(6) Find $\sqrt{\tfrac{7}{11}}$ to 5 places.

$$\tfrac{7}{11} = \cdot\overset{..}{6}\overset{.}{3}$$

```
7  ) ·63 (·79772
      49
149   1463
      1341
1587  12263
      11109
15,9,4 1154
       1116
       38
       32
       6
```

(5) Find $\sqrt{\tfrac{289}{3136}}$.

$$\sqrt{289} = 17 \; ; \quad \sqrt{3136} = 56$$
$$\sqrt{\tfrac{289}{3136}} = \tfrac{17}{56}$$

When the root cannot be expressed exactly, carry the decimal to 6 places.

| | | |
|---|---|---|
| 25. 15·7609 | 35. 11· | 45. ·042849 |
| 26. ·180625 | 36. 45· | 46. ·081 |
| 27. 2889·0625 | 37. 16·675 | 47. $\tfrac{64}{169}$ |
| 28. ·001296 | 38. 28·75 | 48. $\tfrac{224}{361}$ |
| 29. 152·399025 | 39. 43·384675 | 49. $\tfrac{7}{16}$ |
| 30. ·00494209 | 40. 3·16227766 | 50. $\tfrac{3}{4}$ |
| 31. 7· | 41. 7·0030025 | 51. $\tfrac{1}{4}$ |
| 32. 2200· | 42. ·0000016 | 52. $\tfrac{1}{256}$ of $114\tfrac{1}{3}$ |
| 33. ·025 | 43. ·00784 | 53. $\tfrac{1}{125}$ of $48\tfrac{2}{5}$ |
| 34. ·0729 | 44. ·000784 | 54. $2\tfrac{37}{40}$ of $5\tfrac{1}{5}$ |

*The side of a square is found by extracting the square root of its area.*

55. Find the side of a square whose area is 1000 sq. yd.

---

* For conciseness, let $a$ = first part with ciphers having $2n$ or $2n+1$ figures, $b$ = second part with $n-1$ or $n$ figures respectively; then the remainder $= 2ab + b^2$, which divided by the divisor $2a = b + \dfrac{b^2}{2a}$. Now $b^2$ cannot contain more than $2(n-1)$ or $2n$ fig. respectively; and $2a$ not fewer than $2n$ or $2n+1$ respectively; hence $\dfrac{b}{2a}$ is a proper fraction, &c. (See Kelland's Algebra, p. 57.)

**74.** 56. Find the length of the side of a square field containing an acre.

57. The area of Great Britain and Ireland is 122,091 square miles; find the side of a square tract of land of equal extent.

58. How many yd. are in the side of a square, equal in area to a rectangle 972 yd. long and 1296 ft. broad?

59. A rectangle is 240 yd. long and 450 ft. broad; find the side of a square 10 times as large.

60. Find the side of a square of equal extent to 3 fields respectively 15 ac. 3 ro. 17 po.; 11 ac. 3 ro. 36 po.; 5 ac. 1 ro. 36 po.

*Diameters of circles are proportional to the square roots of their areas.*

61. Find the diameter of a circle twice as large as another whose diameter is 120 ft.

62. Find the diameter of a circle $\frac{3}{4}$ of the area of another whose diameter is 30 ft.

*In a right-angled triangle, the square of the hypotenuse is = the sum of the squares of the base and the perpendicular.*

Thus, $AC^2 = AB^2 + BC^2$

(Euclid I. 47).

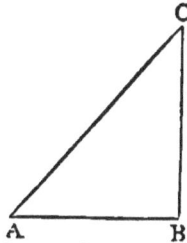

When the *hypotenuse* is *wanted*, we square the base and the perpendicular, and extract the square root of their *sum*. When the *base* or the *perpendicular* is *wanted*, we square the hypotenuse and the perpendicular or the base, and extract the square root of their *difference*.

63. Base = 39, Perpendicular = 52; find Hypotenuse.

64. Base = 180, Perpendicular = 19; find Hypotenuse.

65. Base = 35, Hypotenuse = 91; find Perpendicular.

66. Base = 13, Hypotenuse = 85; find Perpendicular.

67. Perpendicular = 18, Hypotenuse = 82; find Base.

68. Perpendicular = 72, Hypotenuse = 75; find Base.

To obtain *integral* numbers to represent the sides of a right-angled triangle, take any odd number as the base or the perpendicular; from its square, subtract 1, and divide by 2, for the perpendicular or the base; the latter number increased by 1 will be the hypotenuse. Thus, *base = 7*; *perpendicular* $= \frac{49-1}{2} = 24$; *hypotenuse = 25.* Any multiple of these numbers will also suffice.*

---

* Let $n$ = base or perpendicular; $\frac{n^2-1}{2}$ = perpendicular or base; $\frac{n^2+1}{2}$ = hypotenuse. (See *Notes* in Leslie's " Elements of Geometry " on Euclid I., 47.)

**74.**    69. Find the diagonal of a rectangular field whose sides are 20 yd. and 14 yd.

70. Find the diagonal of a wall 28 ft. long and 15 ft. high.

71. Two vessels sail from the same point, the one due north 51 miles, the other due east 68 miles; how many miles are they distant from each other?

72. How many feet from the base of a house must a ladder 27 ft. long be placed to reach a window 21 ft. high?

73. Find the length of a cord stretching from the vane of a steeple 95 ft. high to a point 40 ft. from its base.

74. A cord 287 ft. long is stretched from the top of a column 63 ft. high; find the distance of its point of contact with the ground from the base of the column.

75. A room is 28 ft. long, 21 ft. broad, and 12 ft. high; find the length of the diagonal of the floor or the roof, of the side walls, and of the end walls.

76. In the same room, find the length of the diagonal from a corner of the roof to the opposite corner of the floor.

☞ The square of the diagonal of a room = the sum of the squares of the length, the breadth, and the height; for the sum of the squares of the length and the breadth = square of the diagonal of the floor, which increased by the square of the height = the square of the diagonal of the room.

77. Find the diagonal of a hall, 50 ft. long, 30 ft. broad, and 15 ft. high.

78. Find the breadth of a street from a point in which a ladder 50 ft. long reaches a window 40 ft. high on one side, and another 48 ft. high on the other.

When the same number occupies the 2d and 3d terms of a proportion, it is a *Mean Proportional* between the 1st and 4th. Its square is therefore = the product of the extremes; and the M. P. *of two numbers* is hence = *the square root of their product;* thus, 24 is M. P. of 18 and 32.

79. Find M. P. of 16 and 49.

80. Find M. P. of 84 and 140.

81. Find M. P. of $\frac{7}{12}$ and $\frac{3}{8}$ of $\frac{11}{4}$.

☞ The true weight of a body successively weighed in the scales of a false balance is the M. P. between the apparent weights.*

---

* Let the lengths of the arms of the balance be $a$ and $b$ respectively, $x$ the *true* weight, $m$, $n$, the *apparent* weights.

$$\left. \begin{array}{l} x : m :: a : b \\ n : x :: a : b \end{array} \right\} \therefore x : m :: n : x \text{ and } x = \sqrt{mn}$$

**74.**   82. A body successively weighed in the scales of a false balance appears to be $12\frac{1}{2}$ lb. and $12\frac{1}{4}$ lb. respectively; find its true weight.

83. A body appears to weigh $5\frac{1}{12}$ lb. in one scale and $5\frac{4}{5}$ lb. in the other scale of a false balance; find its true weight.

☞ The times in which bodies fall are proportional to the square roots of the spaces traversed. Since 16·1 ft. is traversed during the *first* second; to find the time, we divide the space by 16·1 and extract the sq. root.

84. In what time will a stone fall to the bottom of a coal pit 70 fathoms deep?

85. In what time would a body fall from the N. or the S. Pole to the centre of the earth, taking the Polar Radius as 20,853,810 ft. ?

---

**75.**               CUBE ROOT.

When the Cube Root of a number is raised to the *third* power, the number itself is reproduced; thus $8 = $ *cube root* of $512 = \sqrt[3]{512}$; $8^3 = 512$.

Take any number, as 59, we know that $59^2 =$ $(50+9)^2 = 50^2 + 2 \times 50 \times 9 + 9^2$. Multiplying by $(50+9)$, we obtain $(50+9)^3$ $= 50^3 + 3 \times 50^2 \times 9 + 3 \times 50 \times 9^2 + 9^3$.

$$
\begin{array}{l}
50^2 + 2(50 \times 9) + 9^2 \\
\underline{50 + 9} \\
50^3 + 2(50^2 \times 9) + (50 \times 9^2) \\
\quad\quad (50^2 \times 9) + 2(50 \times 9^2) + 9^3 \\
\hline
50^3 + 3(50^2 \times 9) + 3(50 \times 9^2) + 9^3
\end{array}
$$

In reproducing 59 or $50+9$, let us determine the method of finding the Cube Root.

$$
\begin{array}{l}
)\,50^3 + 3 \times 50^2 \times 9 + 3 \times 50 \times 9^2 + 9^3\,(50+9 \\
\;\,50^3 \\
\hline
3 \times 50^2 + 3 \times 50 \times 9 + 9^2 \;)\; 3 \times 50^2 \times 9 + 3 \times 50 \times 9^2 + 9^3 \\
\quad\quad\quad\quad\quad\quad\quad\quad\quad\quad\;\; 3 \times 50^2 \times 9 + 3 \times 50 \times 9^2 + 9^3
\end{array}
$$

Subtracting $50^3$, we leave $3 \times 50^2 \times 9 + 3 \times 50 \times 9^2 + 9^3$. Further, to obtain the quotient 9, the divisor must be $3 \times 50^2 + 3 \times 50 \times 9 + 9^2$, or $300 \times 5^2 + 30 \times 5 \times 9 + 9^2$.

(1) Find $\sqrt[3]{205379}$.

A number of *one* figure has no more than *three* figures in its cube; a number of *two* figures has no more than *six*. Since *one* place in a number corresponds to a period of *three* places in its cube, before extracting the cube root, we point off in periods of three places, commencing at units' place.

METHOD I.

$$
\begin{array}{r}
205{,}379(59 \\
125 \\
\hline
80379
\end{array}
$$

$300 \times 5^2 = 7500$
$30 \times 5 \times 9 = 1350$
$9^2 \quad\quad = \;\; 81$

$$
\begin{array}{r}
\overline{8931} \quad\quad 80379 \\
\overline{\hphantom{8931} \quad\quad 80379}
\end{array}
$$

**75.**  The greatest cube root in 205 is 5.  Subtracting $5^3$, we leave 80, which with the next period annexed is 80379.  As we have to add other numbers to $300 \times 5^2 = 7500$, we may require to make repeated trials to obtain the second figure. $7 +$ " some number to be added " may go 9 times in 80.  We then take $30 \times 5 \times 9 = 1350$, and $9^2 = 81$, and adding them to 7500, subtract $9 \times 8931$.  As there is no remainder, we find that $59 = \sqrt[3]{205379}$.  Having thus in the first part subtracted $50^3$, we have next subtracted as much more as makes up $59^3$.

We may vary the *form* of working as in the following methods :—

METHOD II.

$$205379(59$$
$$125$$

$300 \times 5^2 = 7500 \quad \overline{80379}$
$30 \times 5 = 150$
$9$
$9 \times \overline{159} = 1431$
$\overline{8931} \quad 80379$

METHOD III.

$$205379(59$$
$$125$$

$75 \quad \overline{80379}$
$159 \quad 1431$
$\overline{8931} \quad 80379$

In Method II., $9(30 \times 5 + 9) = 30 \times 5 \times 9 + 9^2$.

In Method III., we abridge the process, by *omitting the equivalents*, and, instead of writing ciphers, we merely attend to the relative local value of the figures.

Find the *Cube Root* of the following numbers :—

1.  9261     3.  357911     5.  103823
2.  29791    4.  148877     6.  474552

(2) Find $\sqrt[3]{45499293}$.

METHOD II.

$$45,499,293(357$$
$$27$$

$300 \times 3^2 = 2700 \quad \overline{18499}$
$30 \times 3 = 90$
$5$
$5 \times \overline{95} = 475$
$\overline{3175} \quad 15875$
$300 \times 35^2 = 367500 \quad 2624293$
$30 \times 35 = 1050$
$7$
$7 \times \overline{1057} = 7399$
$\overline{374899} \quad 2624293$

METHOD III.

$$45,499,293(357$$
$$27$$

$27 \quad \overline{18499}$
$95 \quad \left.\begin{array}{l}475\\\overline{3175}\\25\end{array}\right\} \begin{array}{l}15875\\\overline{2624293}\end{array}$
$\overline{3675}$
$1057 \quad 7399$
$\overline{374899} \quad 2624293$

**75.** In METHOD II., having found the first two figures of the root as before, we take $300 \times 35^2$, and finding the third figure to be 7, we make up the divisor as we did for the second figure. In METHOD III., having found the first figure 3, we write $3 \times 3$ or 9 in one column, and $3 \times 9$ or 27 in another. Finding the next figure to be 5, we *annex* 5 to 9, and by putting $5 \times 95$ or 475 *two* places to the right of 27 obtain 3175. By subtracting $5 \times 3175$ from 18499, we find the remainder 2624. We obtain $3 \times 35^2$, by adding $5^2$ or 25 to 3175 and 475. We now triple the last figure of 95, and obtaining 15, write 5 and carry 1 to 9, and thus have $105 = 3 \times 35$. By *annexing* 7 to 105, we add 7 to $30 \times 35$. We now multiply 1057 by 7, and by writing the product *two* places to the right of 3675, we add 7399 or $7(30 \times 35 + 7)$ to 367500 or $300 \times 35^2$. We now subtract $7 \times 374899$, and find $357 = \sqrt[3]{45499293}$.

In the accompanying process we show why $3 \times 35^2$ is obtained by adding $5^2$ to 3175 and 475 :—

$$
\begin{array}{rcl}
2700 &=& 3 \times 30^2 \\
475 &=& \left.\begin{array}{l} 3 \times 30 \times 5 + 5^2 \end{array}\right\} \\
3175 &=& \left.\begin{array}{l} 3 \times 30^2 + 3 \times 30 \times 5 + 5^2 \end{array}\right\} \\
25 &=& 5^2 \\
\hline
3675 &=& 3 \times 30^2 + 6 \times 30 \times 5 + 3 \times 5^2 \\
&=& 3(30^2 + 2 \times 30 \times 5 + 5^2) = 3 \times 35^2.
\end{array}
$$

| | | |
|---|---|---|
| 7. 53157376 | 13. 184608795384 | 19. 570547876184 |
| 8. 62099136 | 14. 103690516392 | 20. 455289041557 |
| 9. 41421736 | 15. 102700479987 | 21. 1881365963625 |
| 10. 12812904 | 16. 305501115375 | 22. 160288833718161 |
| 11. 113379904 | 17. 597585982967 | 23. 184676889190123 |
| 12. 1458274104 | 18. 327510203957 | 24. 497640375631125 |

We may often shorten the operation by Contracted Division.

(3) Find $\sqrt[3]{12396\cdot8834}$.

METHOD III.

$$)12,396\cdot8834(23\cdot14395$$

$$8$$

| | 12 | $\overline{4396}$ |
|---|---|---|
| 63 | 189 } | |
| | $\overline{1389}$ } | 4167 |
| | 9 } | $\overline{229883}$ |
| | $\overline{1587}$ | |
| 691 | 691 } | |
| | $\overline{159391}$ } | 159391 |
| | 1 } | $\overline{70492}$ |
| | $\overline{160083}$ | 64144 |
| 6,9,3 | 277 | 6348 |
| | $\overline{16036,0}$ | 4820 |
| | 28 | $\overline{1528}$ |
| | $\overline{16064}$ | 1446 |
| | 2 | $\overline{82}$ |
| | $\overline{16,0,6,6}$ | |

**75.**   (4) Find $\sqrt[3]{\frac{27}{1331}}$.

$$\sqrt[3]{27} = 3 \qquad \sqrt{1331} = 11$$
$$\sqrt[3]{\tfrac{27}{1331}} = \tfrac{3}{11}$$

(5) Find $\sqrt[3]{\frac{7}{8}}$.

$$\sqrt[3]{\tfrac{7}{8}} = \sqrt[3]{\cdot875}\text{ or} = \tfrac{1}{2}\sqrt[3]{7}$$
$$= \cdot95647-$$

| | | |
|---|---|---|
| 25. 250·047 | 30. ·000970299 | 35. $\frac{64}{729}$ |
| 26. 175·616 | 31. 2126·781656 | 36. $\frac{125}{343}$ |
| 27. 87528·384 | 32. 24212·815957 | 37. $\frac{16}{27}$ |
| 28. ·000068921 | 33. ·00027 | 38. $\frac{8}{27}$ of $\frac{3}{8}$ of $1\frac{5}{16}$ |
| 29. ·000405224 | 34. ·00008 | 39. $\frac{5}{11}$ of $\frac{7}{24}$ of $8\frac{1}{3}$ |

*The side of a cube is found by extracting the cube root of its content or volume.*

40. A cube contains 5832 cub. in.; find the length of its side.

41. The Imperial gallon contains 277·2738 cub. in; find the side of a cube containing a gallon.

42. The *litre*, the French standard of capacity, contains 61·027 cub. in. ; find the side of a cube containing a litre.

*Diameters of spheres are proportional to the cube roots of their contents.*

43. Find the diameter of a sphere *nine* times as large as another whose diameter is 150 ft.

44. The Equatorial Diameter of the Earth is 7926 miles; find that of Venus, whose volume is ·953 of that of the Earth.

*Kepler's Third Law:—the* SQUARES OF THE TIMES *in which the planets revolve round the sun are proportional to the* CUBES OF THEIR MEAN DISTANCES *from the sun.*

45. The periodic time of the Earth is 365·256 da., and of Venus 224·701 da., if the Earth's distance $= 1$, find that of Venus.

$$(365\text{·}256)^2 ; (224\text{·}701)^2 :: 1 : x; \text{ dist. of Venus} = \sqrt[3]{x}.$$

46. The periodic time of Jupiter is 4332·585 da., if the Earth's distance $= 1$, find that of Jupiter.

**76.**   HORNER'S METHOD.

William G. Horner's Method of Finding Roots is applicable to the solution of ANY ROOT.

(1) Find $\sqrt[3]{45499293}$.

| | | |
|---|---|---|
| Having found the greatest | 3 | 9 | 45,499,293(357 |
| cube root in 45 to be 3, we | 3 | 18 | 27 |
| write 3 in one column, $3^2$ or | $\overline{6}$ | $\overline{27}$ | $\overline{18499}$ |
| 9 in another, and subtract $3^3$ | 3 | 475 | 15875 |
| or 27 from 45. We return | $\overline{95}$ | $\overline{3175}$ | $\overline{2624293}$ |
| to the first column, and by | 5 | 500 | 2624293 |
| adding in 3 obtain 6. We | $\overline{100}$ | $\overline{3675}$ | $\overline{\phantom{0}}$ |
| now add $3 \times 6$ or 18 to 9 in | 5 | 7399 | |
| the second column and obtain | $\overline{1057}$ | $\overline{374899}$ | |
| 27. Again, we add 3 to 6 in | | | |
| the first column. | | | |
| Making allowance for what | | | |

**76.** may be carried, we find that 27 when increased may go 5 times in 184. In the first column, we place 5 *one* place to the right of 9 and obtain 95. In the second column, we write 5 × 95 or 475 *two* places to the right, and by adding obtain 3175. 5 × 3175 or 15875 being put *three* places to the right of 18, or under 18499, we obtain the remainder 2624. Returning to the first column, we add 5 to 95. In the second column, we add in 5 × 100 and obtain 3675. In the first column, we again add in 5. Finding the next figure in the root to be 7, we *annex* 7 to 105 in the first column. We place 7 × 1057 *two* places to the right in the second column, and obtaining 374899, place 7 times this sum in the third. We have thus found the CUBE root of 45499293 to be 357. To facilitate comparison, the figures in this process, which are common to the divisors in Methods II. and III. (see page 164), are printed in a *bolder* type.

### (2) Find $\sqrt[4]{12}$.

We place 1, the integral part of the fourth root of 12, in the first column; $1^2$ or 1 in the second; $1^3$ or 1 in the third; and $1^4$ or 1 under 12 in the fourth. In the first column, by adding in 1 to 1 we obtain 2; in the second, by adding in 1 × 2 we obtain 3; and in the third, we add in 1 × 3 and obtain 4. Returning to the first, we add 1 to 2 and obtain 3; and in the second 1 × 3 added to 3 produces 6. We again add in 1 to the first column and obtain 4.

Finding the next figure in the root to be 8, we put 8 *one* place to the right in the first column and obtain 48. We then put 8 × 48 *two* places to the right in the second, and by adding obtain 984. We write 8 × 984 = 7872 *three* places to the right in the third column, and by adding obtain 11872. We then put 8 × 11872 or 94976 *four* places to the right of 11 in the fourth column and subtract it from 110000.

The work is carried on so that while each figure in the FOURTH root is added *four* times in the first column, *three* products are added in the second, *two* are added in the third, and *one* is subtracted in the fourth. After finding the root to be 1·86 we finish the work by Contracted Division.

| | | | |
|---|---|---|---|
| 1 | 1 | 1 | 12 (1·86120972— |
| 1 | 2 | 3 | 1 |
| $\overline{2}$ | $\overline{3}$ | $\overline{4}$ | 110000 |
| 1 | 3 | 7872 | 94976 |
| $\overline{3}$ | $\overline{6}$ | $\overline{11872}$ | $\overline{150240000}$ |
| 1 | 384 | 11456 | 147123216 |
| $\overline{48}$ | $\overline{984}$ | $\overline{23328}$ | $\overline{3116784}$ |
| 8 | 448 | 1192536 | 2576019 • |
| $\overline{56}$ | $\overline{1432}$ | $\overline{24520536}$ | $\overline{540765}$ |
| 8 | 512 | 1218888 | 515702 |
| $\overline{64}$ | $\overline{1944}$ | $\overline{25739424}$ | $\overline{25063}$ |
| 8 | 4356 | 20765 | 23210 |
| $\overline{726}$ | $\overline{198756}$ | $\overline{2576018,9}$ | $\overline{1853}$ |
| 6 | 4392 | 2077 | 1805 |
| $\overline{732}$ | $\overline{203148}$ | $\overline{2578096}$ | $\overline{48}$ |
| 6 | 4428 | 415 | |
| $\overline{738}$ | $\overline{207576}$ | $\overline{257851,1}$ | |
| 6 | 74 | 41 | |
| $\overline{7,4,4}$ | $\overline{20765,0}$ | $\overline{25,7,8,9,2}$ | |
| | 7 | | |
| | $\overline{2,0,7,7,2}$ | | |

**76.** The *fourth root* of a number may be found by taking the square root of its square root; the *sixth root*, by taking the square root of its cube root, &c.

Find the following roots by Horner's Method:

1. $\sqrt[3]{2}$      4. $\sqrt[4]{228886641}$      7. $\sqrt[4]{21224 \cdot 09008801}$

2. $\sqrt[3]{20}$      5. $\sqrt[4]{35806100625}$      8. $\sqrt[6]{81 \cdot 108054012001}$

3. $\sqrt[3]{200}$      6. $\sqrt[4]{20730 \cdot 71593}$      9. $\sqrt[6]{148035889}$

     10. $\sqrt[6]{17}$      11. $\sqrt[4]{\frac{81}{2401}}$      12. $\sqrt[6]{\frac{14}{27} \text{ of } \frac{9}{35} \text{ of } 1\frac{1}{4}}$.

---

**77.**      # SCALES OF NOTATION.

In the common notation, the local value of the figures ascends in the SCALE of TEN. We may, however, adopt other scales: In the scale of 6, " 1 " in the second place being *six* times the value of " 1 " in the first, " 10 " represents 6, the *base* of the scale. Again, " 1 " in the third place being *six* times the value of " 1 " in the second, " 100 " represents 36, the second power of the base. " 2534 " in the scale of 6, or $(2534)_6$, is $= 4 +$ $(3 \times 6) + (5 \times 6^2) + (2 \times 6^3)$; $(65284)_9 = 4 + (8 \times 9) +$ $(2 \times 9^2) + (5 \times 9^3) + (6 \times 9^4)$.

The number of characters used in any scale is denoted by its base. In the scales of 11 and 12, we may represent 10 by D for *Decem;* and in the scale of 12, 11 by U for *Undecim.*

(1) Express 451 in the scale of 6.

$$\begin{array}{r|l} 6 & 451 \\ \hline 6 & 75 \text{ ″} 1 \\ \hline 6 & 12 \text{ ″} 3 \\ \hline & 2 \text{ ″} 0 \end{array}$$

In dividing 451 successively by 6, we find that $451 = (2 \times 6^3) + (0 \times 6^2) + (3 \times 6)$ $+ 1$.

$$451 = (2031)_6.$$

To reduce a number in the decimal scale to its equivalent in another scale, we divide the number successively by the base of the latter, and to the final quotient annex the successive remainders.

1. Red. 666 to scale of 6      4. Red. 313 to scale of 8

2.    ″   315    ″    ″ 4      5.    ″   222    ″    ″ 2

3.    ″   225    ″    ″ 7      6.    ″   1859    ″    ″ 12

(2) Express $(1234)_5$ in the decimal scale.

$$\begin{array}{r} 1234 \\ 5 \\ \hline 7 \\ 5 \\ \hline 38 \\ 5 \\ \hline 194 \end{array}$$

$(1234)_5 = (1 \times 5^3) + (2 \times 5^2) + (3 \times 5) + 4$

$= 5\{(1 \times 5^2) + (2 \times 5) + 3\} + 4$

$= 5\big(5\{(1 \times 5) + 2\} + 3\big) + 4.$

**77.** To reduce a number in any scale to its equivalent in the decimal scale, we multiply the left-hand figure by the base of the former, and add in the next figure to the right, and proceed similarly till all the figures are taken in.

Reduce the following to the decimal scale:

7. $(423)_5$ | 9. $(3567)_8$ | 11. $(2D98)_{11}$
8. $(1243)_6$ | 10. $(12345)_9$ | 12. $(DU10)_{12}$

(3) Express $(2143)_5$ in the scale of 7.

To reduce a number from one scale to another, of which neither is the decimal, we first reduce to the decimal, and then to the required scale.

$$\begin{array}{ll} (2143)_5 & 7\,|\,298 \\ 11 & 7\,\underline{|\,42\,\prime4} \\ \overline{59} & \overline{\quad 6\,\prime0} \\ \overline{298} = (604)_7 \end{array}$$

13. Reduce $(1001001)_2$ to the scale of 3.
14. Reduce $(2D43)_{11}$ to the scale of 7.
15. Reduce $(4U57)_{12}$ to the scale of 2.

The pupil will now see that the *higher* the base of the scale, the *fewer* figures are necessary to represent any number; but if the same number of figures is required in two scales, then the left-hand figure in the *higher* is *less* than that in the *lower* scale.

(4) Reduce 23, 34, 41, to the scale of 3, *add* them and *prove* the work.

$$\begin{array}{ll} 23 = & (212)_3 \qquad (10122)_3 \\ 34 = & (1021)_3 \qquad \overline{3} \\ 41 = & \underline{(1112)_3} \qquad \overline{10} \\ \overline{98} = & (10122)_3 \qquad \overline{32} \\ & \qquad\qquad\qquad \overline{98} \end{array}$$

In adding, we obtain the sum in the 1st column $= 5 = (12)_3$; write 2 and carry 1; in the 2d, $5 = (12)_3$; in the 3d, $4 = (11)_3$; and in the 4th, $3 = (10)_3$.

16. Reduce 64, 127, 95, to the scale of 2, and find the sum.
17. Reduce 23, 143, 79, to the scale of 12, and find the sum.

(5) Reduce 2002 and 1271 to the scale of 4, and find their *difference.*

$$\begin{array}{ll} & \qquad\qquad (23123)_4 \\ 2002 = & (133102)_4 \qquad \overline{11} \\ 1271 = & \underline{(103313)_4} \qquad \overline{45} \\ \overline{731} = & (23123)_4 \qquad \overline{182} \\ & \qquad\qquad\qquad \overline{731} \end{array}$$

3 from 2 we cannot, 3 from 4 leaves 1, 1 and 2 are 3. 1 and 1 make 2, 2 from 4 leaves 2, &c.

H

**77.**    18. Reduce 625 and 367 to the scale of 5, and find their difference.

19. Reduce 237 and 74 to the scale of 9, and find their difference.

The *Arithmetical Complement* (A. C.) of a number in any scale is obtained by subtracting the number from the base, or the next greater power of the base. The Arithmetical Complement of a number is so called because its figures and those of the number *together fill* up the scale.

> In the *decimal* scale, A. C. of $7 = 10 - 7 = 3$; A. C. of $213 = 1000 - 213 = 787$.
>
> In the scale of 6, A. C. of $(3)_6 = (10)_6 - (3)_6 = (6)_6$; A. C. of $(342)_6 = (1000)_6 - (342)_6 = (214)_6$.
>
> The best method of finding A. C. is to commence at the left hand, and subtract each figure from the *base diminished by one*, except the right-hand figure, which we take from the *base*. In the scale of 8, to find A. C. of $(263)_8$, we take 2 from 7, 6 from 7, and 3 from 8, and thus obtain $(515)_8$.

20. In the *decimal* scale, find the A. C. of 43, 726, and 2817.

21. In the scale of 6, find the A. C. of $(24)_6$, $(253)_6$, and $(1243)_6$.

22. In the scale of 12, find the A. C. of $(24)_{12}$, $(346)_{12}$, and $(28DU)_{12}$.

    (6) Reduce 1691 and 127 to the scale of 12, and *multiply* them.

$$
\begin{array}{rcl}
1691 &=& (U8U)_{12} \\
127 &=& (D7)_{12} \\
\hline
11837 & & 6D25 \\
20292 & & 9952 \\
\hline
214757 &=& (D4345)_{12}
\end{array}
\qquad
\begin{array}{r}
(D4345)_{12} \\
124 \\
\hline
1491 \\
17896 \\
214757
\end{array}
$$

7 times $U = 77 = (65)_{12}$.   Write 5 and carry 6, &c.

23. Reduce 2341 and 725 to the scale of 7, and multiply them.

24. Reduce 741 and 1286 to the scale of 6, and multiply them.

25. Reduce 198 and 241 to the scale of 12, and multiply them.

    (7) Reduce 753 and 29 to the scale of 7, and find the quotient.

$$
(41)_7 \,\big)\, (2124)_7 \, \big( (34\tfrac{19}{41})_7 \; = \; 25\tfrac{28}{29}
$$

$$
\begin{array}{r}
153 \\
\hline
264 \\
224 \\
\hline
(40)_7
\end{array}
\qquad \tfrac{763}{779} = 25\tfrac{28}{29}
$$

26. Reduce 864 and 72 to the scale of 3, and find the quotient.

27. Reduce 78467 and 317 to the scale of 12, and find the quotient.

# DUODECIMALS.

**78.**

IN *Duodecimal* Multiplication, we descend in the scale of *twelve* from the *foot*, which is adopted as the *unit* of computation.

A *lineal foot* is divided into 12 *inches* or *primes* ($'$); an inch into 12 *lines*, *parts*, or *seconds* ($''$); a line into 12 *thirds* ($'''$), &c.

Descending from the *square foot* in the duodecimal scale, the names are as follow: *twelfth of sq. ft.* ($'$); *sq. inch* ($''$); *twelfth of sq. in.* ($'''$); *sq. line* ($''''$ or $^{IV}$), &c.

Let AB be a *lineal foot*, divided into 12 in. each $=$ BH. The *square* BD is a *sq. foot*, containing 144 sq. inches, each $=$ BG. BF is the *twelfth of a sq. ft.*, containing 12 sq. in. The twelfth of a sq. ft., which is often *erroneously* called "an inch,"* is a surface, whose length BA is a *foot*, and breadth BE an *inch*.

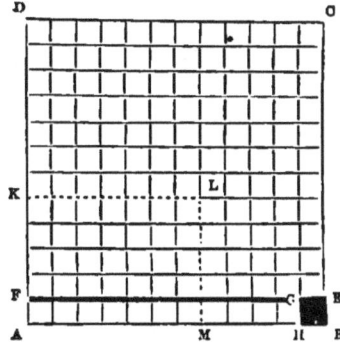

Find the area of the *rectangle* AMLK, *seven* inches long and *five* inches broad. The rectangle contains 5 rows of sq. inches, and in each row there are 7 sq. inches. The number of sq. inches in the rectangle $=$ the product of the number of lineal inches in each dimension $=$ 35 sq. in. We thus see that the product of the *number* of lineal units in the *length* of a surface by the *number* of lineal units in the *breadth* is $=$ the *number* of square units contained in the surface.

(1) Find the area of a surface, 4 ft. 3 in. by 3 ft. 2 in.

$$
\begin{array}{r}
\text{ft. } \;' \\
4 \,''\, 3 \\
3 \,''\, 2 \\
\hline
12 \,''\, 9 \\
8 \,''\, 6 \\
\hline
13 \,''\, 5'\,''\, 6''
\end{array}
$$

$$
\begin{array}{rl}
\text{4 ft. 3 in.} & = \quad 51 \text{ in.} \\
\text{3 ft. 2 in.} & = \quad 38 \text{ in.} \\
\hline
& \quad 408 \\
& \quad 153 \\
\hline
12\,)\,\overline{1938} & \text{ sq. in.} \\
12\,)\,\overline{161'\,''\, 6''} \\
\hline
13\,''\, 5'\,''\, 6''
\end{array}
$$

$$3 \text{ ft.} \times 3 \text{ in.} = 3 \times \tfrac{3}{12} = \tfrac{9}{12}$$
$$3 \text{ ft.} \times 4 \text{ ft.} = \ldots \ldots 12$$
$$2 \text{ in.} \times 3 \text{ in.} = \tfrac{2}{12} \times \tfrac{3}{12} = \tfrac{6}{144}$$
$$2 \text{ in.} \times 4 \text{ ft.} = \tfrac{2}{12} \times 4 = \tfrac{8}{12}$$

Area $=$ 13 sq. ft.; 5 twelfths of sq. ft.; 6 sq. in.

---

* Errors of this kind perpetuated among artificers by their continued use in Ready Reckoners, by confusing names, generate false ideas, of which

**78.**      (2) Find the area of a surface 3 ft. 4′ ″ 7″ long, and 2 ft. 9′ ″ 10″ broad.

<pre>
ft. ′   ″                    ft. ′   ″
3″4 ″ 7                      3″4″ 7  =  487
2″9 ″10                      2″9″10  =  406
───────                             ──────
6″9 ″ 2                              2922
2″6 ″ 5 ″3                           1948
  2 ″ 9 ″9 ″10                      ───────
─────────────                       197722 sq. lines
9″6′″ 5″″0‴″10ᴵⱽ                    ──────
                                    16476″10ᴵⱽ
</pre>

Area = 9 sq. ft., ₁₄₄⁶ sq. ft., 5 <br>
sq. in., ₁₄₄⁹ sq. in., 10 sq. lines.

<pre>
1373″ 0‴
──────
 114″ 5″
──────
   9″ 6′
</pre>

. (3) Find the area of a surface 75 ft. 9′″9″ by 16 ft. 4′″7″.

When the number of feet is greater than 12, we may either keep the number as it is, or *extend the duodecimal scale*. The dimensions = (63)₁₂ ″ 9′ ″ 9″, and (14)₁₂″4′″7″. The product = (875)₁₂ sq. ft., &c. = 1241 sq. ft., &c.

<pre>
6 ″ 3 ″  9′ ″ 9″
1 ″ 4 ″  4 ″ 7
───────────────
6 ″ 3 ″ 9 ″  9
2 ″ 1 ″ 3 ″  3 ″ 0
  2 ″ 1 ″ 3 ″ 3 ″ 0
    3 ″ 8 ″ 2 ″ 8  ″ 3
───────────────────────
8 ″ 7 ″ 5 ″ 11′ ″ 5″ ″ 8‴ ″ 3ᴵⱽ
</pre>

Find the area of surfaces of the following dimensions :—

<pre>
ft. ′    ft. ′        ft. ′  ″   ft. ′  ″      ft. ′   ″    ft. ′    ″
1. 3″ 2× 2″ 3│ 7. 7″1″6× 2″ 4″3│13.28″9″11×11″11″11
2. 5″ 3× 6″ 7│ 8. 4″4″6× 5″ 6″7│14.34″5″ 6×15″ 4″ 7
3. 7″10× 8″11│ 9. 8″9″7× 9″ 6″5│15.43″9″10×28″11″11
4.13″ 6× 9″ 8│10.19″3″6× 7″ 4″9│16.73″6″11×18″ 3″ 6
5.18″ 7× 7″ 8│11.19″8″6×11″10″9│17.64″5″10×16″ 9″ 9
6.18″ 9×12″10│12.32″3″7× 9″11″9│18.76″9″ 5×21″11″ 3
</pre>

19. Find the content of a board 6 ft. 3 in. long, and 4 ft. 7 in. broad.

20. Find the area of a floor 16 ft. 4 in. long, and 14 ft. 8 in. broad.

21. Find the area of a square court whose side is 17 ft. 11 in.

22. What is the content of the ceiling of a square room whose walls are 12 ft. 5½ in. broad?

---

the following is an illustration : A master carpenter once stated that he had often been puzzled by the seeming discrepancy between the extent of a surface as *measured* and as *computed*. Laying down a surface 15 in. by 13 in. he marked off a square foot, and observed that the true content of the remainder seemed to be different from that given in the computed result His difficulty, however, vanished when he found that the answer was NOT 1 sq. ft. 4 in. 3 pts., but 1 sq. ft. 4 *twelfths of sq. ft.*, 3 sq. in. = 1 sq. ft. 51 sq. in. The twelfth of a cubic foot is also erroneously termed an *inch*.

**78.** 23. How much sheet-iron will be required to line the lower half of 12 window shutters, each 8 ft. 2 in. high, and 1 ft. 4 in. broad?

24. How much veneering will be required to cover the surface of 6 counters, of which 2 are each 12 ft. 3 in. by 3 ft. 4 in.; 3 each, 10 ft. 6 in. by 3 ft. 4 in.; and the other 6 ft. 8 in. by 2 ft. 10 in.?

25. How many sq. yards are in the walls of a room, 18 ft. 3 in. in height, and 96 ft. 8 in. in circuit?

26. How many sq. ft. of paper are in a book containing 288 *pages*, each 7 in. by 4¾ in.?

27. How much glass will be required for the front windows of a house of 3 flats: the ground floor containing 6 windows each 7 ft. 4 in. by 3 ft. 4 in., and a fanlight 1 ft. 10 in. by 3 ft. 4 in., and each of the upper flats 7 windows, each 7 ft. by 3 ft. 4 in.?

28. A square court, whose side is 19½ ft., contains a grass-plot 13 ft. 6 in. by 12 ft. 8 in. How much is left to be macadamized?

(4) Find the price of painting a wall 25 ft. 6 in. long, and 14 ft. 4 in. high, @ 1/1½ ℀ sq. yd.

$$
\begin{array}{l}
25''6' \\
14''4' \\
\hline
357''0 \\
8''6''0 \\
\hline
365''6'''0'' \\
9)365\tfrac{1}{2} \text{ sq. ft.} \\
\hline
40\tfrac{11}{18} \text{ sq. yd.}
\end{array}
$$

| | 40 |
|---|---|
| 1½d. ¼ 1/ | 5 |
| ¼ of 1/1½ | 0''8¼ |
| | 45''8¼ |
| | £2''5''8¼ |

29. Find the price of 12 panes of glass, each 1 ft. 5 in. by 11 in @ 2/3 ℀ sq. ft.

30. How much must be paid for lining the bottom of a reservoir 32 ft. 3 in. long, and 14 ft. 8 in. wide, with asphalt @ 2/3 ℀ sq. yd.

31. Find the expense of whitewashing the ceiling of a square room, the breadth of the wall being 10 ft 6 in. @ 3d. ℀ sq. yd.

32. What should be paid for causewaying a street 62 yards long and 12 ft. 6 in. broad, @ 1/6 ℀ sq. yd.?

33. Find the cost of paving a court 58 ft. 9 in. long and 21¾ ft. broad @ 2/3 ℀ sq. yd.?

34. What must be paid for painting a stair of 13 steps, each 2 ft. 7 in. broad, 7 in. high, and 10 in. wide, @ 1/6 ℀ sq. yd.?

(5) Find the superficial content of the walls and ceiling of a room 15 ft. 6 in. long, 12 ft. 4 in. broad, 10 ft. 7 in. high.

**78.**

| ft. ′ | ft. ′ | ft. ′ |
|---|---|---|
| 15 ″ 6 | 55 ″ 8 | 15 ″ 6 |
| +12 ″ 4 | 10 ″ 7 | 12 ″ 4 |
| 27 ″ 10′ | 556 ″ 8 | 186 ″ 0 · |
| 2 | 32 ″ 5 ″8 | 5 ″ 2 ″0 |
| 55 ″ 8′ Circuit. | 589 ″ 1′″8″ Walls. | 191 ″ 2′″0″ Ceiling. |

35. How many sq. ft. arc in the walls of a room 15 ft. 6 in. long, 13 ft. 4 in. broad, and 11 ft. 2 in. high?

36. How many sq. ft. will be required to line a cistern, without lid, 4 ft. 6 in. long, 3 ft. 8 in. broad, and 4 ft. 5 in. deep?

37. Find the cost of painting the walls of a room 13 ft. 6 in. long, 12 ft. broad, and 9 ft. high, @ 1/6 ℣ sq. yard.

38. Find the cost of painting the outside of a box, except the bottom, length and breadth each 3 ft. 4 in., and depth 2 ft. 8 in., @ 1/3 ℣ sq. yd.

39. How many sq. yds. of plastering arc in the walls and ceiling of a room in the form of a cube 12 ft. each way, deducting for window 6 ft. 3 in. by 3 ft. 2 in., door 7 ft. 6 in. by 3 ft. 6 in., and fireplace 4 ft. 3 in. by 3 ft. 4 in.?

40. How many copies of a pictorial newspaper of 4 *pages*, each 28 in. by 20 in., will be required to cover the walls of a country barber's shop, 18 ft. 8 in. long, 14 ft. 4 in. broad, and 8 ft. 10 in. high, allowing for 2 windows each 5 ft. 6 in. by 3 ft. 2 in.; 2 doors each 7 ft. by $3\frac{1}{2}$ ft.; and fireplace 3 ft. by $2\frac{1}{2}$ ft.?

Descending from the *cubic foot* in the duodecimal scale, the names are: $\frac{1}{12}$ of cub. ft. (′), $\frac{1}{144}$ of cub. ft. (″), cub. in. (′″), $\frac{1}{12}$ of cub. in. ($^{IV}$), $\frac{1}{144}$ of cub. in. ($^{V}$), cub. line ($^{VI}$).

(6) Find the cubic content of a solid, 11 ft. $4\frac{1}{3}$ in. long, 3 ft. $3\frac{5}{12}$ in. broad, and 2 ft. $4\frac{1}{4}$ in. thick.

| 11 ″ 4′″4″ | | 37″ 3′″9″″9′″ ″8$^{IV}$ |
|---|---|---|
| 3 ″ 3 ″5 | | 2″ 4 ″3 |
| 34 ″ 1 ″0 | | 74″ 7 ″7 ″ 7 ″4 |
| 2 ″10 ″1 ″0 | | 12″ 5 ″3 ″ 3 ″2 ″8 |
| 4 ″8 ″9 ″8 | | 9 ″3 ″11 ″5 ″5 ″0 |
| 37 ″ 3′″9″″9′″″8$^{IV}$ | | 87″10′″2″″10′″″0$^{IV}$″1$^{V}$″0$^{VI}$ |

Cubic content = $(87 + \frac{10}{12} + \frac{2}{144})$ cub. ft. + $(10 + \frac{0}{12} + \frac{1}{144})$ cub. in. + 0 cub. lin.

Find the cubic content of solids of the following dimensions.—

| ft. ′ ″ | ft. ′ ″ | ft. ′ ″ | | ft. ′ ″ | ft. ′ ″ | ft. ′ ″ |
|---|---|---|---|---|---|---|
| 41. 8″11 × | 7″8 × | 6″7 | 44. | 11″3″4 × | 6″9″10 × | 5″4″6 |
| 42. 9″ 6 × | 6″6 × | 4″3 | 45. | 12″4″6 × | 8″6″ 8 × | 4″6″6 |
| 43. 9″ 7 × | 6″8 × | 5″4 | 46. | 9″6″7 × | 3″4″ 5 × | 5″4″3 |

78. 47. Find the solidity of a block of granite 8 ft. 4 in. long, 6 ft. 6 in. broad, 5 ft. 7 in. thick.

48. Find the cubic content of a slab of marble 5 ft. 6 in. long, 4 ft. 3 in. broad, 1 ft. 10 in. thick.

49. How many cubic ft. of air are in a room 35 ft. 6 in. long, 20 ft. 8 in. broad, and 12 ft. 4 in. high?

50. Find the weight of sea-water in a cistern 11 ft. 3 in. long, 6 ft. 7 in. broad, and 5 ft. 6 in. deep, the weight of a cubic foot of sea-water being 1025 oz.

51. Find the weight of a log of oak 10 ft. 5 in. long, and 2 ft. 3 in. square throughout, the weight of a cubic foot of oak being 925 oz.

52. Find the cost of a block of lead 1 ft. 3 in. long, 9 in. broad, $8\frac{1}{2}$ in. thick, taking the weight of a cubic foot of lead at 709 lb. and the price @ £23„10 ℔ ton.

---

# SERIES.

A SERIES is a succession of numbers which mutually depend on one another, according to a certain law.

## 79. ARITHMETICAL PROGRESSION.

An ARITHMETICAL PROGRESSION (A. P.) is a series of numbers uniformly *ascending* or *descending* by a constant *difference*, and is therefore appropriately termed an EQUIDIFFERENT SERIES.*

1, 4, 7, 10, 13, &c., is an ascending A. P., in which 1 is the *first term*, and 3 the *common difference*. 50, 46, 42, 38, 34, &c., is a descending A. P., in which 50 is the *first term*, and 4 the *common difference*.

| The former series is as follows: | | | The latter series is as follows: | | |
|---|---|---|---|---|---|
| Term | | | Term | | |
| 1st | $= 1$ | $= 1$ | 1st | $= 50$ | $= 50$ |
| 2d | $= 1 + (1\times3)$ | $= 4$ | 2d | $= 50 - (1\times4)$ | $= 46$ |
| 3d | $= 1 + (2\times3)$ | $= 7$ | 3d | $= 50 - (2\times4)$ | $= 42$ |
| 4th | $= 1 + (3\times3)$ | $= 10$ | 4th | $= 50 - (3\times4)$ | $= 38$ |
| 5th | $= 1 + (4\times3)$ | $= 13$ | 5th | $= 50 - (4\times4)$ | $= 34$ |
| &c. | &c. | &c. | &c. | &c. | &c. |

---

* A HARMONICAL PROGRESSION (H. P.) is a series of numbers which are respectively the reciprocals of the terms of an Arithmetical Progression; thus, 3, 5, 7, 9, are in A. P., and $\frac{1}{3}$, $\frac{1}{5}$, $\frac{1}{7}$, $\frac{1}{9}$, in H. P. $\frac{2}{3}$, $\frac{1}{2}$, $\frac{1}{3}$, $\frac{1}{6}$, are in A. P., and $\frac{3}{2}$, 2, 3, 6, in H. P.

**79.** To obtain any term in an A. P., we multiply the common difference by the number less by one than the number showing the rank of the term in the series, and add the product to the first term, or subtract it from it, according as the series is ascending or descending.

In the first series, the 100th term is $= 1 + (99 \times 3) = 298$.
In the second, the 10th term is $= 50 - (9 \times 4) = 14$.

(1) Find the 36th term in the A. P. 5, $5\frac{1}{2}$, &c.
Difference $= \frac{1}{2}$; 36th term $= 5 + (35 \times \frac{1}{2}) = 22\frac{1}{2}$.

(2) Find the 20th term in the A. P. 7, $6\frac{7}{8}$, &c.
Diff. $= \frac{1}{8}$; 20th term $= 7 - (19 \times \frac{1}{8}) = 4\frac{5}{8}$.

Find the
1. 10th term in 1, 3, 5, &c.
2. 100th  „  2, 4, 6, &c.
3. 25th  „  7, 11, 15, &c.
4. 73d  „  18, 22, 26, &c.
5. 36th  „  $1\frac{1}{2}$, $2\frac{1}{4}$, 3, &c.

Find the
6. 13th term in $3\frac{3}{4}$, $4\frac{1}{8}$, $4\frac{1}{2}$, &c.
7. 100th  „  ·015, ·02, ·025, &c.
8. 50th  „  100, $99\frac{1}{2}$, 99, &c.
9. 30th  „  50, $48\frac{3}{4}$, $47\frac{1}{2}$, &c.
10. 19th  „  12, 11·75, 11·5, &c.

11. A number of nuts is divided among 30 boys. The first gets 120, and each boy gets 3 fewer than the one preceding. How many does the thirtieth get?

12. A clerk is engaged for £70 the first year, with an increase of 7 guineas for every successive year. Find his salary for the seventh year.

13. A body falls 16·1 feet during the first second; thrice as far during the second; five times during the third; and so on. How far would a body fall during the sixth second?

14. Of seven frigates, the first has 66 guns, the second has 4 fewer, and so on with the same difference. How many has the seventh?

15. Thirteen trucks are laden with coal; the first contains 5·65 tons, and each truck has 2·5 cwt. more than the one preceding. How much coal is on the last truck?

Take the A. P. 8, 11, 14, 17, 20, 23.
We find that  $8 + 23 = 31$
 $11 + 20 = 31$
 $14 + 17 = 31$
Sum of the A. P. $= 3(8+23) = 3 \times 31$.

Take the A. P. 70, 63, 56, 49, 42.
We find that  $70 + 42 = 112$
 $63 + 49 = 112$
 $56 = \frac{1}{2}$ of 112
Sum of the A. P. $= \frac{5}{2}(70+42) = 2\frac{1}{2} \times 112$.

**79.** The sum of an A. P. is = the product of the sum of the first and the last term by half the number of terms.*

Any term in an A. P. is the *Arithmetical Mean* between two terms *equidistant* from it; thus, 14 is the A. M. between 11 and 17; 17, the A. M. between 11 and 23; 56, the A. M. between 70 and 42.

(3) Find the sum of the series 2, 5, to 51 terms.

51st term $= 2 + (50 \times 3) = 152.$

$$S = \tfrac{51}{2}(2 + 152) = \frac{51 \times 154}{2} = 3927.$$

16. Find S. of 4, 10, 16, to 50 terms
17.  "   "  $\frac{1}{2}, \frac{3}{4}, 1,$  " 30  "
18.  "   "  $\frac{2}{3}, \frac{15}{18}, 1,$ " 40  "

19. Find S. of ·01, ·03, ·05, to 29 terms
20.  "   "  2, 1·9, 1·8, " 15  "
21.  "   "  80, 77$\frac{1}{2}$, 75, " 30  "

22. In Venice the clocks strike to 24. How many strokes are made in a day?

23. A boy gains 10 marbles on Monday, 3 more on Tuesday, and 3 more on each successive day. How many has he gained in six days?

24. A merchant gained £90 during the first year in business, and £35 more in each successive year than the one preceding. How much has he gained in 20 years?

25. A labourer saved 1d. the first week of the year, and $\frac{1}{4}$d. more on each successive week. How much has he at the end of the year?

26. A body falls 16·1 ft. during the first second, thrice as far during the second, and so on. How far would a body fall in six seconds?

27. If 20 sentinels are placed in a line at the successive distance of 40 yards; how far will a person travel who goes from the 1st to the 2d and back; from the 1st to the 3d and back; and so on till he goes from the 1st to the 20th and back : and how long will he take at the average rate of 3$\frac{1}{2}$ miles ℣ hour?

---

**80.** GEOMETRICAL PROGRESSION.

A GEOMETRICAL PROGRESSION (G. P.) is a series of numbers uniformly *ascending* or *descending* by a common *ratio;* and is therefore appropriately termed an EQUIRATIONAL SERIES.

2, 6, 18, 54, &c., is an ascending G. P., in which the *common ratio* is $\frac{6}{2}$ or 3.  1, $\frac{1}{2}$, $\frac{1}{4}$, $\frac{1}{8}$, &c., is a descending G. P., in which the *common ratio* is $\frac{1}{2}$.

---

* Let $a =$ the first term, $d =$ the difference, S $=$ the sum, and $l =$ the nth term, or the *last* of n terms; then $l$ or the nth term $= a \pm (n-1)d,$

$$S = \tfrac{n}{2}(a + l) = \tfrac{n}{2}\{2a \pm (n-1)d\}.$$

H 2

**80.** The former series is as follows:

| Term | | |
|---|---|---|
| 1st | $= 2$ | $= 2$ |
| 2d | $= 2 \times 3$ | $= 6$ |
| 3d | $= 2 \times 3^2$ | $= 18$ |
| 4th | $= 2 \times 3^3$ | $= 54$ |
| &c. | &c. | &c. |

The latter series is as follows:

| Term | | |
|---|---|---|
| 1st | $= 1$ | $= 1$ |
| 2d | $= 1 \times \frac{1}{2}$ | $= \frac{1}{2}$ |
| 3d | $= 1 \times (\frac{1}{2})^2$ | $= \frac{1}{4}$ |
| 4th | $= 1 \times (\frac{1}{2})^3$ | $= \frac{1}{8}$ |
| &c. | &c. | &c. |

To obtain any term in a G. P., we raise the common ratio to the power whose index is less by one than the number showing the rank of the term in the series, and then multiply the power by the first term.

In the 1st series, the 11th term is $= 2 \times 3^{10} = 118098$.

In the 2d series, the 20th term is $= 1 \times (\frac{1}{2})^{19} = \frac{1}{524288}$.

(1) Find the 9th term in the G. P. 7, 21, &c.

Ratio $= \frac{21}{7} = 3$; 9th term $= 7 \times 3^8 = 45927$.

(2) Find the 6th term in the G. P. $2\frac{1}{4}$, $1\frac{1}{2}$, &c.

Ratio $= 1\frac{1}{2} \div 2\frac{1}{4} = \frac{2}{3}$; 6th term $= 2\frac{1}{4} \times (\frac{2}{3})^5$
$$= \frac{9}{4} \times \frac{32}{243} = \frac{8}{27}.$$

Find the
1. 6th term in 4, 8, 16, &c.
2. 5th     "    7, 28, 112, &c.
3. 9th     "    $\frac{1}{5}$, 1, 5, &c.

Find the
4. 10th term in 81, 27, 9, &c.
5. 7th     "    $\frac{8}{6}$, $\frac{3}{4}$, $\frac{1}{2}$, &c.
6. 5th     "    $\frac{9}{10}$, $\frac{8}{100}$, $\frac{6}{1000}$, &c.

7. Of seven purses, the first contains 1/4; the second, 2/; the third, 3/; and so on in the same ratio. How much does the last contain?

8. A person who found a potato imitated the example of Samuel Budgett and planted it. At the end of the first season he obtained 25 potatoes; and during each successive season the whole crop of the preceding one was planted and increased in the same ratio. Find the crop at the end of the fifth season.

9. Out of a vessel containing 10 gallons of brandy, $\frac{1}{5}$ was extracted and replaced with water, $\frac{1}{5}$ of the content was again extracted and replaced with water, and so on for seven times. How much brandy is finally in the vessel?

☞ The first term is 10, the ratio $\frac{4}{5}$, and the number of terms 8.

Let us find the sum of the G. P. 2, 6, 18, 54.

$$\text{Ratio} = \frac{6}{2} = 3.$$

$1 \times \text{Sum} = 2 + (2 \times 3) + (2 \times 3^2) + (2 \times 3^3)$

$3 \times \text{Sum} = \quad (2 \times 3) + (2 \times 3^2) + (2 \times 3^3) + (2 \times 3^4)$

$(3 - 1) \text{ Sum} = (2 \times 3^4) - 2 = 2 (3^4 - 1)$

$$\text{Sum} = 2 \times \frac{3^4 - 1}{2} = 80.$$

**80.** Let us find the sum of the G. P. 9, $1\frac{2}{7}$, $\frac{9}{49}$, $\frac{9}{343}$.

Ratio $= 1\frac{2}{7} \div 9 = \frac{1}{7}$.

$1 \times \text{Sum} = 9 + (9 \times \frac{1}{7}) + \{9 \times (\frac{1}{7})^2\} + \{9 \times (\frac{1}{7})^3\}$

$\frac{1}{7} \times \text{Sum} = \quad (9 \times \frac{1}{7}) + \{9 \times (\frac{1}{7})^2\} + \{9 \times (\frac{1}{7})^3\} + \{9 \times (\frac{1}{7})^4\}$

$(1 - \frac{1}{7})\,\text{Sum} = 9 - \{9 \times (\frac{1}{7})^4\} = 9\{1 - (\frac{1}{7})^4\}$

$$\text{Sum} = 9 \times \frac{1 - (\frac{1}{7})^4}{1 - \frac{1}{7}} = 10\frac{170}{343}.$$

To find the sum of a G. P. we raise the ratio to the power denoted by the number of terms, divide the difference between this power and unity by the difference between the ratio and unity, and multiply the quotient by the first term.*

Any term in a G. P. is a Mean Proportional, or a Geometrical Mean between two *equidistant* terms; thus, in the G. P. 16, 24, 36, 54, 81; $36 = \sqrt{16 \times 81} = \sqrt{24 \times 54}$ (see § 57 & § 74.)

(3) Find the sum of 7, 14, 28, to 10 terms.

Ratio $= \frac{14}{7} = 2$. Sum $= 7 \times \frac{2^{10} - 1}{2 - 1} = 7161$.

(4) Find the sum of $\frac{1}{7}$, $\frac{1}{28}$, $\frac{1}{112}$, to 8 terms.

Ratio $= \frac{1}{28} \div \frac{1}{7} = \frac{1}{4}$.

$$\text{Sum} = \frac{1}{7} \times \frac{1 - (\frac{1}{4})^8}{1 - \frac{1}{4}} = \frac{1}{7} \times \frac{65535}{65536} \times \frac{4}{3} = \frac{21845}{114688}.$$

|   |   |   | terms. |   |   |   | terms. |
|---|---|---|---|---|---|---|---|
| 10. | Find the sum of | 2, 4, 8, | to 12 | 13. | Find the sum of 3, $\frac{3}{4}$, $\frac{3}{16}$, to 7 |
| 11. | " | " | 5, 15, 45, " 8 | 14. | " | " | $\frac{5}{27}$, $\frac{5}{9}$, $\frac{5}{3}$, " 5 |
| 12. | " | " | $\frac{1}{6}$, $\frac{5}{16}$, $\frac{25}{32}$, " 10 | 15. | " | " | $\frac{2}{15}$, $\frac{2}{3}$, $\frac{2}{135}$, " 6 |

16. Of seven boys, the first has 64 nuts, the second 96, and so on in the same ratio. How much have they in all?

17. Of five brothers, the eldest has £759·7·6, the second two-thirds of this sum, and so on in the same ratio. How much have they in all?

18. A gentleman on taking a house for twelve months ignorantly agreed to pay 1 mil as rent for the first month, 1 cent for the second, 1 florin for the third, and so on in the same ratio. To what would the rent amount?

The number of terms in a descending G. P. may sometimes be infinite; thus every Interminate Decimal is an infinite descending G. P.

---

* Let $a =$ the first term, $r =$ common ratio, $S =$ the sum, $l =$ the last of $n$ terms; then $l$ or the $n$th term $= ar^{n-1}$,

$$S = a\left(\frac{r^n - 1}{r - 1}\right) = \frac{ar^n - a}{r - 1} = \frac{rl - a}{r - 1}.$$

**80.** In $\cdot\dot{7}$, which is $= \frac{7}{10} + \frac{7}{100} + \frac{7}{1000} + $ &c. *ad infin.* ($\infty$), the ratio is $\frac{1}{10}$.

Now a fraction when raised to a power becomes less as the index of the power becomes greater; when therefore the index is infinite, the fraction becomes 0.

Hence, Sum which is $= \frac{7}{10} \times \dfrac{1 - (\frac{1}{10})^{\infty}}{1 - \frac{1}{10}}$ is $= \frac{7}{10} \times \dfrac{1 - 0}{1 - \frac{1}{10}}$

$$= \frac{\frac{7}{10}}{1 - \frac{1}{10}} = \frac{\frac{7}{10}}{\frac{9}{10}} = \frac{7}{9}.$$

The sum of an Infinite descending G. P. is $=$ the first term divided by the difference between the ratio and unity.*

(5) Find the sum of $\frac{185}{1000} + \frac{185}{1000000} + $ &c.$^{\infty}$

Ratio $= \frac{1}{1000}$; Sum $= \frac{185}{1000} \div (1 - \frac{1}{1000}) = \frac{185}{999}$.
See § 34, No. 1.

19. Find the value of $\cdot\dot{4}\dot{5}$, or the sum of $\frac{45}{100} + \frac{45}{10000} + $ &c.$^{\infty}$

20. Find the value of $\cdot\dot{0}3\dot{7}$, or the sum of $\frac{37}{1000} + \frac{37}{1000000} + $ &c.$^{\infty}$

---

**81.**                 COMPOUND INTEREST.

WHEN a sum is lent for a number of periods or terms at COMPOUND INTEREST, the Interest is added to the Principal at the end of each term, and the Amount obtained becomes the Principal for the next term.

On £600 lent for 5 years @ 5 %, the *Simple Interest* would be £150; and the Amount, £750. But at *Compound Interest* the Amount would be as follows:—

| | |
|---|---:|
| Principal for the *first* year . . . . | £600 |
| Interest      *"*      *"*      *"*   . . . . | 30 |
| Principal for the *second* year . . . . | 630 |
| Interest      *"*      *"*      *"*   . . . . | 31·5 |
| Principal for the *third* year . . . . . | 661·5 |
| Interest      *"*      *"*      *"*   . . . . | 33·075 |
| Principal for the *fourth* year . . . . | 694·575 |
| Interest      *"*      *"*      *"*   . . . . | 34·72875 |
| Principal for the *fifth* year . . . . . | 729·30375 |
| Interest      *"*      *"*      *"*   . . . . | 36·4651875 |
| Amount for 5 years   . . . . . | 765·7689375 |
| Original Principal   . . . . . | 600 |
| Compound Interest   . . . . . | £165·7689375 |

---

\* When $n$ is infinite, and $r < 1$. $r^n = 0$, $S = \dfrac{a}{1 - r}$.

**81.**  Exercises in Compound Interest may be performed by this method, but a more concise plan may be obtained by considering the following :—

$$\text{Interest on £1 for 1 year @ 5 °/}_o = \overset{£}{\cdot 05}$$
$$\text{Amount on £1 } \textit{ " " " " " } = 1\cdot 05$$

Since the Amount for any year becomes the Principal for the next, we obtain the following proportions :—

| Principal. | | Amount. | | |
|---|---|---|---|---|
| £ | £ | £ | £ | |
| $1 : 1\cdot 05$ | $: : 1\cdot 05$ | $: 1\cdot 05^2$ | $=$ Am$^t$ for 2 years. | |
| $1 : 1\cdot 05^2$ | $: : 1\cdot 05$ | $: 1\cdot 05^3$ | $=$ | $"$ $"$ 3 $"$ |
| $1 : 1\cdot 05^3$ | $: : 1\cdot 05$ | $: 1\cdot 05^4$ | $=$ | $"$ $"$ 4 $"$ |
| $1 : 1\cdot 05^4$ | $: : 1\cdot 05$ | $: 1\cdot 05^5$ | $=$ | $"$ $"$ 5 $"$ |

Therefore $1 : 600 \quad : : 1\cdot 05^5 : 600 \times 1\cdot 05^5 = £765\cdot 7689375$

To find the AMOUNT of a given sum for a number of terms at Compound Interest, we raise the Amount of £1 for one term to the power denoted by the number of terms, and multiply by the given sum.

(1) Find the Amount of £450 and the Compound Interest on it for 3 years @ 4 °/$_o$.

Am$^t$ of £1 for 1 yr.  @ 4 °/$_o$ = £1·04
Am$^t$ of £450 for 3 yr. @ 4 °/$_o$ = 450 × 1·04$^3$
= 450 × 1·124864 = £506·189 = £506$"$3$"$9¼
Comp$^d$ Int. = £506$"$3$"$9¼ — £450 = £56$"$3$"$9¼.

In involving the Amount of £1, we take as many places in the powers as will produce the result correct to *three* decimal places* (see § 39, § 73, & § 43.)

(2) Find the Amount of £547·625 for 4 years @ 5 °/$_o$, payable half-yearly.

Am$^t$ of £1 for ½ yr. @ 5°/$_o$ ℀ ann.=£1·025
Am$^t$ of £547·625 for 8 half years @ 5 °/$_o$ = 547·625
× 1·025$^8$ = £667·228

---

* Calculations in Compound Interest are often effected by having the amounts of £1 at the important rates tabulated for a series of years. Exercises in Compound Interest afford good illustration of the advantages of Logarithms. The Questions prescribed above are, however, given for such periods as enable them to be easily solved by Involution.

**81.**          Find the Amount of the following sums :—

| | |
|---|---|
| 1. £600 for 2 years @ 3 % | 7. £697„15„0 for 6 yrs. @ 2¼ % |
| 2.   300 „ 3  „      „ 5 % | 8.  468„10„6 „ 4  „    „ 4⅝ % |
| 3.   800 „ 4  „      „ 3 % | 9.  232„ 7„6 „ 8  „    „ 3 % |
| 4.  .400 „ 4  „      „ 4 % | 10.   35„ 3„9 „ 3  „    „ 3¼ % |
| 5.   700 „ 4  „      „ 2¼ % | 11. 666„13„4 „ 5  „    „ 2½ % |
| 6.   834 „ 5  „      „ 3½ % | 12. 267„19„2 „ 7  „    „ 4¼ % |

13. Find the Amount of £670 for 3 years @ 6 %, supposing the interest to become due half-yearly.

14. Find the Amount of £684 for 3 years @ 4 %, supposing the interest to be due quarterly.

15. What is the Compound Interest on £764„12„6 for 4 years @ 5 %, due half-yearly?

16. Find the Compound Interest on £29„15 for 3½ years @ 3½ %, due quarterly.

17. Find the difference between the Simple and the Compound Interest on £750 for 3 years @ 4½ %.

18. A sum of £300 is lent for one year @ 4 %; find the difference between the Simple and the Compound Interest, due quarterly.

19. To what will a legacy of £500 left to a boy 11 years of age have accumulated at Compound Interest, on his attaining *majority* at 21 years of age, allowing Interest @ 5 %?

20. A legacy of £2500 was left to a young lady in 1852 on condition that it should be improved at Compound Interest for a marriage-portion. To what will it have accumulated at her marriage in 1860, reckoning Interest @ 5 %?

We may require to find the Principal which, improved at Compound Interest, may at a future date amount to a given sum; thus, let us find a sum which in 6 years @ 3½ % will amount to £700..

Am$^t$. of £1 for the given time $= £1 \cdot 035^6 = £1 \cdot 229255$

$$\underset{\text{Amount.}}{£1 \cdot 229255} : £700 :: \underset{\text{Principal.}}{£1} : x = £\frac{700}{1 \cdot 229255}$$

We work by Contracted Division (see § 40.), and obtain the result £569·450, the Present Value of £700.

To find the PRESENT VALUE of a given sum due in a given time at a given rate, we divide the given sum by the amount of £1 for the given time.*

---

\* Let P = Principal, A = Amount, R = Rate, $n =$ number of years,

$$A = P \left(1 + \frac{R}{100}\right)^n; \quad P = \frac{A}{(1 + \frac{R}{100})^n}$$

**81.**	(3) Find the Present Value of £500$_{\prime\prime}$12$_{\prime\prime}$6, due in 7 years at 3%.

$$1\cdot03^7 = 1\cdot229874$$

$$\pounds\frac{500\cdot625}{1\cdot229874} = \pounds407\cdot054 = \pounds407_{\prime\prime}1_{\prime\prime}1.$$

Find the Present Value of

21.£900 due in 2 yrs. @ 4% | 24.£1405$_{\prime\prime}$11$_{\prime\prime}$6 due in 4 yr.@4%
22. 700 $_{\prime\prime}$ $_{\prime\prime}$ 4 $_{\prime\prime}$ $_{\prime\prime}$ 5% | 25. 105$_{\prime\prime}$11$_{\prime\prime}$3 $_{\prime\prime}$ $_{\prime\prime}$ 3 $_{\prime\prime}$ $_{\prime\prime}$ 3$\frac{1}{4}$%
23.1200 $_{\prime\prime}$ $_{\prime\prime}$ 4 $_{\prime\prime}$ $_{\prime\prime}$ 3% | 26. 333$_{\prime\prime}$ 3$_{\prime\prime}$4 $_{\prime\prime}$ $_{\prime\prime}$ 5 $_{\prime\prime}$ $_{\prime\prime}$ 2$\frac{1}{4}$%

27. What sum will in 3 years @ 4 % amount to £100, supposing the interest to be paid quarterly?

28. Find the sum which, with half-yearly payments of interest, will at 6 % amount in 4 years to £253·354.

29. A merchant who has increased each year's capital by a *tenth*$_{\prime\prime}$4 finds that at the end of twelve years he has £3985$_{\prime\prime}$16$_{\prime\prime}$1$\frac{1}{2}$. Find his original capital.

30. A sloop was bought by A, who sold it to B, by whom it was sold to C, who finally disposed of it. Each gained 30 % on his prime cost. C sold it for £659$_{\prime\prime}$2; what did A pay for it?

One of the most important applications of *Compound Interest* is in the calculation of ANNUITIES. An Annuity, as its name imports, is a sum payable yearly for a certain number of years; an Annuity may, however, be payable at equal intervals of any duration, as *half-yearly, quarterly,* &c.

Suppose a person, entitled to an annuity of £30 ℘ annum for 5 years, payable yearly, draws none of it till the end of the time; to what will it have amounted, reckoning interest at 4 %?

£1 of the annuity might be lent at the first payment for 4 years, and become at Compound Interest £1·04⁴; £1 at the second payment might be lent for 3 years, and become £1·04³; £1 at the third payment might be lent for 2 years, and become £1·04²; £1 at the fourth payment would in 1 year become £1·04; and to these we would add the fifth payment of £1.

The Amount of an Annuity of £1 for 5 years @ 4 % is thus $= \pounds1\cdot04^4 + 1\cdot04^3 + 1\cdot04^2 + 1\cdot04 + 1$. The sum of this Geometrical Progression (see § 80.) is $= \pounds\frac{1\cdot04^5-1}{1\cdot04-1} = \pounds\frac{1\cdot04^5-1}{\cdot04}$. Having found the Amount of an Annuity of £1, that of £30 for the same time and rate $= \pounds30 \times \frac{1\cdot04^5-1}{\cdot04}$.

To find the AMOUNT of an Annuity, we diminish the amount of £1 for the given time and rate by £1, divide the differ-

**81.** ence by the interest of £1 for one term, and multiply the quotient by the given Annuity.

> (4) Find the Amount of an Annuity of £25 payable half-yearly in 4 years @ 5 %.
>
> Int. of £100 for $\frac{1}{2}$ yr.$=$£2·5 ; Int. of £1 for 1 hf.yr.$=$£·025
> Am$^t$ of £1 for 8 hf.yr.$=$£1·025$^8$ ; Annuity for $\frac{1}{2}$ yr.$=$£12·5
>
> Amount of Annuity $= £12·5 \times \frac{1·025^8-1}{·025} = £12·5 \times \frac{·2184029}{·025}$
> $= £12·5 \times 8·736116 = £109·20145.$

31. Find the amount of an annual rent of £25 for 8 years @ 5 %.

32. Find the amount of an annuity of £60 payable yearly for 6 years @ $3\frac{1}{2}$ %.

33. The Lord Justice Clerk of Scotland has an annual salary of £4500. To what would it amount in seven years @ 4 % ?

34. Find the amount of an annuity of £36 payable quarterly for $2\frac{1}{4}$ years @ $2\frac{1}{2}$ %.

35. A gentleman of fortune, entitled to an annual pension of £200, payable half-yearly, allows it to accumulate for 10 years. Find the amount @ 5 %.

36. A salary of £180, payable quarterly, is not drawn for $1\frac{3}{4}$ years. Find the amount @ 5 %.

37. Find the amount of 4 half-yearly dividends of £2000 stock in the three per cents, reckoning interest @ 4 %.

     ☞ The half-yearly annuity is *one-half* of 3 % on £2000.

Suppose a person, desirous of obtaining an annuity of £70 ℞ annum for 10 years, wishes to know how much he must pay for it @ 3 %.

The amount of this would be $£70 \times \frac{1·03^{10}-1}{·03}$. The sum to be paid for the annuity would evidently be that which in 10 years would produce this amount. We would therefore require to find the Present Value of the Amount by dividing it by the amount of £1 for the given time (see p. 183).

$$£70 \times \frac{1·03^{10}-1}{·03} \div 1·03^{10} = £70 \times \frac{1-\frac{1}{1·03^{10}}}{·03}.$$

To find the PRESENT VALUE of an Annuity, we diminish £1 by the Present Value of £1 for the given time and rate, divide the difference by the interest of £1 for one term, and multiply the quotient by the given Annuity.

> (5) Find the Present Value of an Annuity of £30„17„6 payable quarterly in $2\frac{3}{4}$ years @ $3\frac{1}{2}$ %.

**81.** Int. of £100 for $\frac{1}{4}$ yr. $= £·875$; Int. of £1 for 1 quar. $= £·00875$

Present Value of £1 for 11 quarters $= £\frac{1}{1·00875^{11}}$ ;

Annuity for 1 qr. $= £7·71875 = \frac{1}{4}$ of £30·875.

Present Value of Annuity $= £7·71875 \times \dfrac{1 - \frac{1}{1·00875^{11}}}{·00875}$

$\qquad = £7·71875 \times \dfrac{1 - ·908618}{·00875} = £7·71875 \times 10·4436$

$\qquad = £80·6115.$

When the time is unlimited, the Annuity is termed a PERPETUITY.*

A person wishing to obtain a perpetuity of £200 ⅌ annum is desirous of knowing the sum to be paid for it @ 5%.
The amount of any sum, as £1, for an unlimited time being ∞ (infinite), its reciprocal, or the present value of £1, due in an unlimited time, is hence $= 0$. Present Value of £1 $= £\frac{1}{1·05^{\infty}} = 0$. Present Value of Perpetuity $= £200 \times \dfrac{1-0}{·05} = £\frac{200}{·05} = £4000$.

The sum of £4000 lent out @ 5 % will produce £200 in perpetuity.

(6) Find the Present Value of a Perpetuity of £99₍₎2₍₎6 ⅌ annum @ 3$\frac{1}{4}$ %.

$$\text{Present Value} = £\frac{99·125}{·0325} = 3050.$$

38. Find the present value of an annuity of £40 payable annually for 10 years @ 4$\frac{1}{2}$ %.

39. Find the present value of an annuity of £62₍₎10 for 3$\frac{1}{2}$ years, payable half-yearly, @ 5%.

40. Find the present value of a perpetuity of £210₍₎17₍₎6 ⅌ annum @ 3$\frac{1}{4}$ %.

41. The Lord Justice General of Scotland has a salary of £4800 ⅌ annum. Find the present value of this for 10 years @ 3%.

42. A tenant, on taking a lease of a house for 7 years @ £19 ⅌ annum, pays the present value. Find the sum, reckoning interest @ 4 %.

---

\* Let $a =$ Annuity, R=Rate %, $r = \frac{R}{100} =$ Int. on £1, $n =$ N° of years,

Amount of an Annuity $\qquad = a \times \dfrac{(1+r)^{n} - 1}{r}$.

Present Value of an Annuity $= a \times \dfrac{1 - \frac{1}{(1+r)^{n}}}{r}$.

Present Value of a Perpetuity $= \dfrac{a}{r}$ or $\dfrac{100a}{R}$.

**81.** 43. A colonel of the Royal Marines on half-pay has £264⸳12⸳6 ℣ annum. Find the present value of this annual salary for 6 years @ 4%.

44. What ought to be paid for a property giving an annual rent of £187⸳8⸳6, reckoning @ 4½ %?

45. What sum paid in January 1858 will produce an annuity of £50, payable half-yearly until July 1861, @ 4½ %?

---

## MISCELLANEOUS EXERCISES.

**82.**

1. FIND the L. C. M. of all the multiples of 3 from 6 to 27 inclusive.
2. Find the G. C. M. of $25 \times 45$ and $5 \times 3^5$.
3. What is the G. C. M. of the square of 48 and the cube of 18 ?
4. Find the L. C. M. of the first ten even numbers.
5. Reduce $\frac{7}{9}\frac{4}{9}\frac{5}{9}\frac{7}{9}\frac{1}{9}\frac{5}{9}$ to its lowest terms.
6. Arrange $\frac{3}{4}$, $\frac{23}{35}$, $\frac{43}{55}$, $\frac{11}{14}$, and $\frac{11}{17}$, in order of magnitude.
7. Subtract the sum of $\frac{1}{3} + \frac{3}{8} + \frac{5}{8} + 1\frac{1}{2} + 1\frac{3}{4}$ from 5.
8. Find that number of which $(\frac{5}{6} + \frac{3}{4} - \frac{7}{8})$ is $= 51$.
9. Multiply $\frac{7}{8}$ of $2\frac{1}{3}$ by $\dfrac{2\frac{2}{3} + \frac{3}{4}}{2\frac{3}{4} - \frac{3}{4}}$.
10. Multiply $\frac{7}{8} + \frac{5}{6}$ by $\frac{7}{8} - \frac{5}{6}$, and increase the product by $\frac{6}{11}$ of $1\frac{3}{5}$.
11. Find that number whose fifth diminished by its seventh is $= 3\frac{3}{4}$.
12. From the square root of ·000169 subtract the square root of ·00016.
13. Find the decimal which when added to the difference of $2\frac{5}{6}$ and ·002775 produces the square of ·215.
14. Subtract the cube of 1·6 from 130 times ·0325.
15. Find the interest on £·219 for 47 days @ 3·6 %.
16. A grocer by selling sugar @ 6½d. ℣ lb. loses ¼d.; find his loss %.
17. From Edinburgh to Glasgow by railway is 47½ miles. In what time will a train traverse the line at the rate of 990 yards ℣ min., allowing $\frac{5}{24}$ hour for stoppages?
18. The number of copies in the first edition of the *Lay of the Last Minstrel*, which was 750, was to that in the seventh as 15 to 71. Find the number in the latter.
19. From 1847 to 1857, the Revenue of the City of Edinburgh was £70,629, and the Expenditure £57,684. What per-centage was the difference or surplus of the former?
20. A, at the rate of 4¼ miles an hour, walks a distance in $3\frac{1}{10}$ hours; in what time will B walk the same distance at the rate of $\frac{4}{5}$ of 5¼ miles an hour?

**82.**  21. Find the square root of $10^5 - 316^2$.

22. Find the cube root of $\cdot 296$.

23. Find the H. P. of an engine which can raise $4\frac{1}{2}$ tons of coals per hour from a pit 77 fathoms deep.

24. The centre arch in Westminster Bridge, which is 76 feet wide, is the seventh from the side, and each arch is 4 ft. narrower than the adjoining one nearer the centre. Find the width of the first arch.

25. Divide £5 among A, B, C, D, in the mutual ratios of $\frac{1}{2}$, $\frac{1}{6}$, $\frac{1}{8}$, and $\frac{1}{3}$.

26. A sum of £1343.14.6 collected for a family of orphans was laid out at 6 % per annum. Find the value of a half-yearly payment.

27. Reduce 1 dwt. to the decimal of 1 lb. avoir.

28. Divide 25832 in the ratios of the squares of the reciprocals of the first four odd numbers.

29. The Admirals, the Vice Admirals, and the Rear Admirals of the British Navy are each divided into 3 classes of Red, White, and Blue, and the classes of each rank contain the same number. The number of Admirals is $\frac{7}{3}$ of that of the whole, which is 99, and that of Vice-Admirals is $\frac{3}{7}$. Find the number in each class of Admirals.

30. The walls of Rome erected by Aurelian have been calculated to contain $1396\frac{1}{2}$ hectares, each $2\cdot47114$ acres. Express the area in sq. miles.

31. The weight of an American dollar is $412\frac{1}{2}$ grains, of which $\frac{9}{10}$ is pure silver. Find the weight of pure silver in 100 dollars.

32. If on every guinea of selling price half-a-crown is gained; find the gain on £1000 of buying price.

33. Victoria Bridge on the St Lawrence is within 50 yards of 2 miles in length; in what time will a train traverse it at the rate of $\frac{5}{8}$ of 36 miles in $\frac{2}{3}$ hour?

34. Find the difference between the Compound Interest on £840 for 4 years @ 5 %, and for 5 years @ 4 %, payable half-yearly in each case.

35. The Horticultural Society, founded by Sir Joseph Banks in 1804, was remodelled in 1856; during $\frac{3}{4} + \frac{2}{13}$ of this period it had been incorporated by Royal Charter, in what year did it receive the charter?

36. How many days elapsed between the Annular Eclipse of 15th May 1836 and that of 15th March 1858?

37. Find the Simple and the Compound Interest on £180 for $3\frac{1}{2}$ years @ 5 %.

38. Express the cube of 11 in the scale of 3.

39. From York to London is a distance of 192 miles; two trains

**82.** start at the same time from each terminus, the one from York at the rate of 40 and the other at 32 miles an hour. How far from London will they meet?

40. F starts at $12^h$ at $6\frac{7}{8}$ miles an hour, and B at $12^h$ $30^m$. At what rate must B travel to overtake F at $2^h$?

41. 3 Russian versts are $= 3500$ yards. Reduce a verst to the decimal of a mile.

42. Assuming the length of a glacier, described by Principal Forbes, to be 20 miles, and its annual progression 500 ft.; how long would a block of stone take to traverse its length?

43. Of 150 encumbered estates in Ireland, the numbers in the four provinces were respectively as 1, 2, 3, and 4. Find the number in each province.

44. How many metres are in a Scotch mile, taking 1 Scotch mile $= 1\cdot123024$ Imperial mile; and 1 metre $= 39\cdot37079$ inches.

45. How many metres are in a Scotch mile, taking the following approximations, 8 Scotch miles $= 9$ Imperial miles, and 32 metres $= 35$ yards?

46. If, in victualling a crew, 80 days are allowed for an outward and homeward voyage to Oporto; $\frac{5}{4}$ of this time for one to Demerara; to Boston, $\frac{1}{4}$ of that to Demerara; to Valparaiso, $\frac{9}{5}$ of that to Boston. Find the time allotted for an outward and homeward voyage to Valparaiso.

47. In 1851, the population of Glasgow was $3\cdot58866$ per cent. of the population of Edinburgh more than double the latter, which was 161,648. Find the former.

48. Of an estate, the uncultivated part is $\frac{2}{15}$, the cultivated part, $\frac{3}{4}$, and the remainder under wood contains 65 acres. How many acres are in the whole?

49. In 1855, in the naval armament of France, the number of line of battle ships was $\frac{3}{4}$ of 100; that of frigates, which was 12 less than $\frac{3}{4}$ of the number of the line of battle ships, was $\frac{7}{8}$ of that of the smaller vessels, and the number of steam vessels was $\frac{33}{8}$ of double the number of frigates. How many were there of each?

50. What length of rails requires 873 T. 1 cwt. 1 qr. for their construction @ $\frac{5}{8}$ cwt. $\psi$ yard?

51. Find the greatest depth of Lake Erie, which is $\frac{2}{5}$ of that of Lake Huron, whose greatest depth is $\frac{3}{5}$ of that of Lake Ontario, which is $\frac{5}{8}$ of that of Lake Michigan, whose greatest depth is $\frac{10}{9}$ of that of Lake Superior, which is 990 ft.

52. In the Walcheren Expedition, out of an average force ot 40,589, there were 4212 deaths. Find the per-centage.

53. The circulation of a periodical was 38,500; of the whole, the

**82.** number of stamped copies was $1\frac{3}{4}$. How many copies were unstamped?

54. Find the surface of a floor 28 ft. $7\frac{1}{2}$ in. long, and 15 ft. $6\frac{3}{4}$ in. broad.

55. Find the sum of $\sqrt{1\frac{1}{2}} + \sqrt[3]{1\frac{9}{2}} + \sqrt[4]{2\frac{1}{3}\frac{5}{5}}$.

56. Find that number whose square root is $= \frac{2}{3}$ of $5\frac{1}{4} + \frac{3}{3}$ of $1\frac{1}{3}$.

57. Find the true discount on £22″17″3¾ for 3 months at $5\frac{1}{2}$ %.

58. In 1850, the states of Ohio and Tennessee, nearly of equal extent, produced 59,078,695 and 52,276,223 bushels of wheat respectively. Find the difference of their weight, reckoning the bushel at $\frac{11}{16}$ cwt.

59. If A pays $11\frac{1}{2}$d. ⅌ £ for income-tax, what is his income when the net proceeds are £116″3″1?

60. If a courier traversed a distance of 400 miles in 36 hours, in what time did he traverse $\frac{2}{3}$ of $\frac{3}{8}$ of $\frac{9}{7}$ of $6\frac{3}{4}$ miles?

61. No. 1585 of the *Athenæum* appeared on 13th March 1858; on the hypothesis that it has been regularly published once a-week, find the date of No. 1.

62. Find the price of 3 cwt. 2 qr. 13 lb. carrots @ 16/ ⅌ 240 lb.

63. Reduce a talent of 3000 shekels, each $\frac{1}{2}$ oz. avoir., to the decimal of 1 cwt.

64. A train contains 13 trucks laden with coals; the average weight of a loaded truck is 10 T. 8 cwt. 1 qr., and that of an empty truck 3 T. 16 cwt. 2 qr. Find the weight of coals conveyed by the train.

65. If, in the Russian tariff, the duty on Scotch herrings is 40 copecs ⅌ pood; how much sterling is this ⅌ cwt., a ruble of 100 copecs being $= 3/1\frac{1}{2}$, and a pood being $= 36$ lb. avoir.?

66. A miser collected £370 in packets of pound notes, crowns, half-crowns, florins, shillings, and sixpences. The values of five of the packets were the following fractions of the whole:—packet of notes, $\frac{8 6}{195}$; of crowns, $\frac{13}{370}$; of hf.-crowns, $\frac{11 1}{4 8 5}$; of florins, $\frac{1}{18}$; of shillings, $\frac{1}{50}$. Find the number of notes, crowns, hf.-crowns, florins, shillings, and sixpences.

67. Find the value of $\frac{2}{3}$ cr. $+ \frac{3}{4}$ s. $- \frac{5}{8}$ fl. $+ £\frac{3}{4}$.

68. A can do a work in $7\frac{1}{2}$ days, B in $6\frac{3}{4}$ days, and C in $5\frac{5}{8}$ days. In what time can they do it by working together?

69. A field contains 18 ac. 2 ro. 18 po., and another 7 ac. 3 ro. 7 po. Find the side of a square field of equal area to both.

70. Of the two members of parliament returned for Manchester in 1857, the number of votes polled by one amounted to 8118, and was to its excess above the other number as 902 to 47. Find the latter number.

71. The circulation of a newspaper in the first quarter of a year

**82.** was 3200, and in the second quarter 3600. What would the circulation in the third quarter require to be to show the same ratio of increase?

72. An angler, by using a single hook and a tackle of four hooks alternately for equal times during a day, caught 9½ lb. with the former, and 11 lb. with the latter. On another day he caught 25 lb. with the former; what might he have taken with the latter?

73. Texas contains 274,362 square miles. Into how many lots, each 4536 acres, might it be divided?

74. The managers of a congregation buy a site of ⅔ rood for £500. How much will they pay for ₇⁄₁₀ acre?

75. A person who has paid £6″4″2 of income-tax has £142″15″10 over. How much has he paid ₩ £?

76. 70 masons can build a mansion in 61 days; after working for 10 days, 15 more are engaged. How many days fewer will be occupied than would otherwise have been?

77. If 7 men can do as much as 11 youths, and if 21 youths can do a work in 13 days; in what time can 14 men and 4 youths do it?

78. A starts on a journey at the rate of 3½ miles an hour, B follows in ⅜ hour at the rate of 4 miles an hour. How far on will B overtake A?

79. In the household book of a ducal family we have the following entry by the steward:—"Given your lordship on New Year's Day to give your grandchildren and the servants and several others, £32″6″6." Taking this as Scots money, which is *one-twelfth* of sterling, express the sum as the fraction of £100 sterling.

80. Find the 10th term of the series 1, 1½, 2, &c.

81. Find the 10th term of the series 1, 1½, 2¼, &c.

82. The population of a country in 1854 was 4,500,000, and if it has increased each year at the rate of 10 per cent. on the preceding year; find the population in 1859.

83. In an estate in Sweden, the arable land contains 200 tunnlands; meadowland, 2 per cent. less than the arable; and woodland, 1⅞ per cent. less than 7 times the arable. Find the area in acres of the estate, a tunnland being = 1·2312 acre.

84. From Dresden to Prague by rail is 150 kilometres, each 1093·63 yards: a train leaves Dresden at the rate of 48 kilometres ⴖ hour, and in a quarter of an hour afterwards another leaves Prague at the rate of 40 kilometres ⴖ hour. How many miles from Prague will they meet?

85. Give eight convergents to the fraction which a kilometre is of a mile.

, 86. What sum invested in the 3½ per cents at 93¾ will produce £61″5?

87. Deposited £500 in the National Bank on 1st September 1856, when interest was @ 2½ %; on 8th Oct. it rose to 3½ %, and on 15th May to 4 %. Find the interest due on 8th June 1857.

88. At what rate must £273 be lent from 1st January to 27th May to produce £4·914 of interest?

89. Find the price of 19 cwt. 2 qr. 7 lb. @ £1″8″6 ℣ cwt. by decimals.

90. Find the value of 17·375 cwt. @ £·5625 by decimals and by practice.

91. In the reign of Henry VIII., among the monasteries and religious houses whose revenues were confiscated, there were 186 belonging to the Benedictines with a revenue of £65,879″14, and 173 to the Augustines with a revenue of £18,691″12″6. Reduce the average of one of the latter to the decimal of that of one of the former.

92. Divide the square of 390,404,646 by the square of 123,456,789, and let the quotient be carried out till it contains 3 significant figures in the decimal.

93. A capitalist who had invested £3120 sterling in stock @ 97½, sold £2500 stock @ 98, and the remainder of the stock @ 96. Find his gain or loss.

94. A labourer's wages for 30 days are £3″18″9. Find the wages for the working days in January and February 1860, new year's day being on a Sunday.

95. If 1 lb. troy of sterling gold is worth £46⅞⅜, find the weight of 3465 sovereigns and 1792 hf.-sovereigns.

96. The Brisbane Prize Fund of the Royal Scottish Society of Arts amounts to £175 in the 3 per cent. Government Consolidated Annuities. Find the value of the fund at 90; and the value of the prize.

97. Find the weight of an oaken block 2·25 ft. long, 16 inches broad, and $\frac{9}{13}$ of 1·625 ft. thick; a cubic foot of water weighing 999·278 oz. avoir., and the specific gravity of oak being ·925.

98. The deflection of the Earth's curvature is 8 inches for 1 mile, 32 inches for 2 miles, and so on, the deflection being proportional to the square of the distance. Find the height of a light above the level of the sea which is visible for 14 nautical miles, each 6076 ft.

99. 42 men, whose average strength is ⅘ of that of an ordinary man, can do a piece of work in 4½ days which other 54 men can do in 4⅞ days. What is the average strength of one of the latter set as compared with that of an ordinary man?

100. The period of the Earth's revolution is 365·256 days, and

**82.** that of Mercury 87·969 days.  Express by Kepler's Law the decimal that the distance of Mercury from the Sun is of that of the Earth.

101. In a heavy gale, a flagstaff 60 ft. high snaps 28·8 ft. from the bottom, and not being wholly broken off, the top touches the ground.  How far is its point of contact from the bottom?

102. Seventeen trees are standing in a line 20 yards apart from each other; a person walks from the first to the second and back, thence to the third and back, and so on to the end.  How far will he have walked?

103. If the value of 1 oz. troy of sterling gold $\frac{1}{1}\frac{1}{2}$ fine is £3·89375; find the value of 1 lb. avoir of pure gold.

104. A lunation = 29·53 days, is the period in which the moon passes once through her phases.  After a cycle of 223 lunations, known to the Chaldeans as *Saros*, eclipses recur in the same order and magnitude.  Find the date of the eclipse in the next cycle corresponding to the solar eclipse of 28th July 1851.

105. When a body floats in a liquid the weight of the liquid displaced is = the weight of the floating body.  The effective length of a vessel is 96 ft., the effective breadth 22½ ft., and the draught of water 9 ft.  Find the weight of the vessel, taking the weight of a cubic foot of water roughly at 62½ lb.

106. Find the mean discharge per second of the River Tay, supposing the area of its basin to be 2400 square miles, the annual fall of rain to be 30 inches, of which ⅛ is lost in evaporation.

**THE END.**

PRINTED BY OLIVER AND BOYD, EDINBURGH.

# EDUCATIONAL WORKS

PUBLISHED BY

## OLIVER AND BOYD, EDINBURGH;

SOLD ALSO BY

### SIMPKIN, MARSHALL, AND CO., LONDON.

*\*\** *A Specimen Copy of any work will be sent to Principals of Schools, post free, on receipt of one half the retail price in postage stamps.*

## Lessons from Dr M'Culloch's First Reading-Book,
printed with LARGE TYPE, in a Series of Ten Sheets, for Hanging on the Wall. Price 1s. ; or mounted on Roller, 1s. 8d.

## Dr M'Culloch's Manual of English Grammar, Philosophical and Practical ; with Exercises ; adapted to the Analytical mode of Tuition. 1s. 6d.

## English Prefixes and Affixes. 2d.

In all the books of Dr M'Culloch's series, the important object of exercising the juvenile mind by means of lessons on useful and interesting subjects is steadily kept in view. Directions are given relative to the mode of teaching, as well as tables and lists calculated to assist in the process of instruction. On this point the *Spectator* newspaper, when reviewing the series, remarked :—" In recommending these books, it must not be conceived that we recommend them as likely to save trouble to the teacher, or to operate by witchcraft on the pupil. At their first introduction they will require some care on the part of the master, as well as the exercise of some patience, to enable the pupil to profit by the lessons. But this once done, their foundation is sound, and their progress sure. And let both parents and teachers bear in mind that these are the only means to acquire real knowledge."

## Poetical Reading-Book, with Aids for Grammatical
Analysis, Paraphrase, and Criticism ; and an Appendix on English Versification. By J. D. MORELL, A.M., LL.D., Author of Grammar of the English Language, etc., and W. IHNE, Ph. D. 2s. 6d. Containing,

| | |
|---|---|
| THE DESERTED VILLAGE. | THE MERCHANT OF VENICE. |
| THE TASK (Book I.) | MISCELLANEOUS SELECTIONS. |
| PARADISE LOST (Books I. and V.) | THE PRISONER OF CHILLON. |
| THE FIELD OF WATERLOO. | |

Dr Morell, in the preface to his "Grammar of the English Language," says—" As great care was taken to adapt this book [the Poetical Reading-Book] to the requirements of teachers using the Grammar, and special marks invented for indicating the correct analysis of the poetical extracts contained in it, I take the present opportunity of recommending it to the attention of the higher classes of schools in the country."

## English Grammar, founded on the Philosophy of Language
and the Practice of the best Authors. With Copious Exercises, Constructive and Analytical. By C. W. CONNON, LL.D. 2s. 6d.

*Spectator.*—" It exhibits great ability, combining practical skill with philosophical views."

## Connon's First Spelling Book. 6d.

## Outlines of English Grammar and Analysis, for
ELEMENTARY SCHOOLS, with EXERCISES. By WALTER SCOTT
DALGLEISH, M.A. Edin., one of the Masters in the London Intor-
national College. 8d. KEY, 1s.

*Preface.*—"Aims at providing a COMMON-SCHOOL GRAMMAR which shall be
fully abreast of the latest developments of the science, and at the same time
thoroughly practical and simple in its mode of treating the subject."

## Dalgleish's Progressive English Grammar, with EXER-
CISES. 2s. KEY, 2s. 6d.

*From* Dr JOSEPH BOSWORTH, *Professor of Anglo-Saxon in the University of
Oxford; Author of the Anglo-Saxon Dictionary, etc., etc.*

"Quite a practical work, and contains a vast quantity of important information, well arranged, and brought up to the present improved state of philo-
logy. I have never seen so much matter brought together in so short a space."

## Dalgleish's Grammatical Analysis, with PROGRESSIVE
EXERCISES. 9d. KEY, 2s.

## Dalgleish's Introductory Text-Book of English
COMPOSITION, based on GRAMMATICAL SYNTHESIS; containing
Sentences, Paragraphs, and Short Essays. 1s.

## Dalgleish's Advanced Text-Book of English Com-
POSITION, treating of Style, Prose Themes, and Versification.
2s. Both Books bound together, 2s. 6d. KEY, 2s. 6d.

## A Dictionary of the English Language, containing
the Pronunciation, Etymology, and Explanation of all Words author-
ized by Eminent Writers. By ALEXANDER REID, LL.D., late
Head Master of the Edinburgh Institution. Reduced to 5s.

The Work is adapted to the present state of the English language and the
improved methods of teaching. While the alphabetical arrangement is pre-
served, the words are grouped in such a manner as to show their etymological
affinity; and after the first word of each group is given the root from which
they are derived. These roots are afterwards arranged into a vocabulary.
At the end is a Vocabulary of Classical and Scriptural Proper Names.

## Dr Reid's Rudiments of English Grammar. 6d.
## Dr Reid's Rudiments of English Composition. 2s.
KEY, 2s. 6d.

The volume is divided into three parts: Part I. is meant to guide to correctness in spelling, punctuation, the use of words, and the structure and arrangement of sentences; Part II. to correctness and perspicuity in style, and to a
tasteful use of ornament in writing; and Part III. to the practice of the
preceding rules and exercises in various kinds of original composition.

**Lennie's Principles of English Grammar.** Comprising the Substance of all the most approved English Grammars, briefly defined, and neatly arranged; with Copious Exercises in Parsing and Syntax. *Revised Edition;* with the author's latest improvements, and an Appendix in which Analysis of Sentences is fully treated. 1s. 6d.

**The Author's Key,** containing, besides Additional Exercises in Parsing and Syntax, many useful Critical Remarks, Hints, and Observations, and *explicit and detailed instructions as to the best method of teaching Grammar.* 3s. 6d.

**Analysis of Sentences;** being the Appendix to Lennie's Grammar adapted for General Use. Price 3d.—KEY, 6d.

**The Principles of English Grammar;** with a Series of Progressive Exercises. By Dr JAMES DOUGLAS, lately Teacher of English, Great King Street, Edinburgh. 1s. 6d.

**Douglas's Initiatory Grammar,** intended as an Introduction to the above. 6d.

**Douglas's Progressive English Reader.** A New Series of English Reading Books. *The Earlier Books are illustrated with numerous Engravings.*

| | | |
|---|---|---|
| FIRST BOOK. 2d. | THIRD BOOK. 1s. | FIFTH BOOK. 2s. |
| SECOND BOOK. 4d. | FOURTH BOOK. 1s. 6d. | SIXTH BOOK. 2s. 6d. |

**Douglas's Selections for Recitation,** with Introductory and Explanatory Notes; for Schools. 1s. 6d.

**Douglas's Spelling and Dictation Exercises.** 144 pages, price 1s.

*Athenæum.*—"A good practical book, from which correct spelling and pronunciation may be acquired."

**Shakspeare's King Richard II.** With Historical and Critical Introductions; Grammatical, Philological, and other Notes, etc. Adapted for Training Colleges. By Rev. Canon ROBINSON, M.A., late Principal of the Diocesan Training College, York. 2s.

**Wordsworth's Excursion. The Wanderer.** With Notes to aid in Analysis and Paraphrasing. By Canon ROBINSON. 8d.

**English Composition for the Use of Schools.** By ROBERT ARMSTRONG, Madras College, St Andrews; and THOMAS ARMSTRONG, Heriot Foundation School, Edinburgh. Part I., 1s. 6d. Part II., 2s. Both Parts bound together, 3s. KEY, 2s.

**Armstrong's English Etymology.** 2s.

**Armstrong's Etymology for Junior Classes.** 4d.

## History of English Literature; with an OUTLINE of the ORIGIN and GROWTH of the ENGLISH LANGUAGE. Illustrated by EXTRACTS. For Schools and Private Students. By WILLIAM SPALDING, A.M., Professor of Logic, Rhetoric, and Metaphysics, in the University of St Andrews. *Continued to* 1870. 3s. 6d.

*Spectator.*—"A compilation and text-book of a very superior kind. . . . Mr Spalding has brought to his survey not only a knowledge of our history and literature, but original reflection, a comprehensive mind, and an elevation of tone, which impart interest to his account, as well as soundness to his decision. The volume is the best introduction to the subject we have met with."

*Athenæum.*—"The numerous extracts scattered throughout the volume are well chosen for the purpose of throwing light on the authors from whom they are taken, and at the same time exhibiting the gradual advance of our literature from its earliest to its present state. Mr Spalding's critical remarks are discriminating, impartial, judicious, and always well put."

## Selections from Paradise Lost; with NOTES adapted for Elementary Schools, by Rev. ROBERT DEMAUS, M.A., F.E.I.S., late of the West End Academy, Aberdeen. 1s. 6d.

The Selections have been chosen so as to furnish a continuous narrative.

## Demaus's Analysis of Sentences. 3d.

## System of English Grammar, and the Principles of Composition. With numerous Exercises, progressively arranged. By JOHN WHITE, F.E.I.S. 1s. 6d.

## Millen's Initiatory English Grammar. 1s.

## Ewing's Principles of Elocution, improved by F. B. CALVERT, A.M. 3s. 6d.

Consists of numerous rules, observations, and exercises on pronunciation, pauses, inflections, accent, and emphasis, accompanied with copious extracts in prose and poetry.

## Rhetorical Readings for Schools. By WM. M'DOWALL, late Inspector of the Heriot Schools, Edinburgh. 2s. 6d.

## Object-Lesson Cards on the Vegetable Kingdom. Set of Twenty in a Box. £1, 1s.

The design of this Series is to give a short description of some Plants which are cultivated for their useful properties, *each subject being illustrated with specimens* (attached to the Cards) *of the various objects described*, and forming in this department an interesting Industrial Museum, which will be found of great value in the education of the young.

## How to Train Young Eyes and Ears; being a MANUAL of OBJECT-LESSONS for PARENTS and TEACHERS. By MARY ANNE ROSS, Mistress of the Church of Scotland Normal Infant School, Edinburgh. 1s. 6d.

**Household Economy:** a MANUAL intended for Female Training Colleges, and the Senior Classes of Girls' Schools. By MARGARET MARIA GORDON (Miss Brewster), Author of " Work, or Plenty to do and how to do it," etc. 2s.

*Athenæum.*—" Written in a plain, genial, attractive manner, and constituting, in the best sense of the word, a practical domestic manual."

### SESSIONAL SCHOOL BOOKS.

**Etymological Guide.** 2s. 6d.

This is a collection, alphabetically arranged, of the principal roots, affixes, and prefixes, with their derivatives and compounds.

**Old Testament Biography,** containing notices of the chief persons in Holy Scripture, in the form of Questions, with references to Scripture for the Answers. 6d.

**New Testament Biography,** on the same Plan. 6d.

**Fisher's Assembly's Shorter Catechism Explained.** 2s.

PART I. Of what Man is to believe concerning God.
II. Of what duty God requires of Man.

# GEOGRAPHY AND ASTRONOMY.

IN compiling the works on these subjects the utmost possible care has been taken to ensure clearness and accuracy of statement. Each edition is scrupulously revised as it passes through the press, so that the works may be confidently relied on as containing the latest information accessible at the time of publication.

**School Geography.** By JAMES CLYDE, LL.D., one of the Classical Masters of the Edinburgh Academy. With special Chapters on Mathematical and Physical Geography, and Technological Appendix. *Corrected throughout.* 4s.

In composing the present work, the author's object has been, not to dissect the several countries of the world, and then label their dead limbs, but to depict each country, as made by God and modified by man, so that the relations between the country and its inhabitants—in other words, the present geographical life of the country—may appear.

*Athenæum.*—" We have been struck with the ability and value of this work, which is a great advance upon previous Geographic Manuals. . . . Almost for the first time, we have here met with a School Geography that is quite a readable book,—one that, being intended for advanced pupils, is well adapted to make them study the subject with a degree of interest they have never yet felt in it. . . . Students preparing for the recently instituted University and Civil Service examinations will find this their best guide."

## Dr Clyde's Elementary Geography. *Corrected throughout.* 1s. 6d.

In the *Elementary Geography* (intended for less advanced pupils), it has been endeavoured to reproduce that life-like grouping of facts—geographical portraiture, as it may be called—which has been remarked with approbation in the *School Geography.*

## A Compendium of Modern Geography, POLITICAL,

PHYSICAL, and MATHEMATICAL : With a Chapter on the Ancient Geography of Palestine, Outlines of Astronomy and of Geology, a Glossary of Geographical Names, Descriptive and Pronouncing Tables, Questions for Examination, etc. By the Rev. ALEX. STEWART, LL.D. *Carefully Revised.* With 11 Maps. 3s. 6d.

## Geography of the British Empire. By WILLIAM

LAWSON, St Mark's College, Chelsea. *Carefully Revised.* 3s.

PART I. Outlines of Mathematical and Physical Geography. II. Physical, Political, and Commercial Geography of the British Islands. III. Physical, Political, and Commercial Geography of the British Colonies.

## Edinburgh Academy Modern Geography. *Carefully Revised.* 2s. 6d.

## Edinburgh Academy Ancient Geography. 3s.

## An Abstract of General Geography, comprehending a

more minute description of the British Empire, and of Palestine or the Holy Land, etc. With numerous Exercises. For Junior Classes. By JOHN WHITE, F.E.I.S., late Teacher, Edinburgh. *Carefully Revised.* 1s. ; or with Four Maps, 1s. 3d.

## White's System of Modern Geography; with Outlines of

ASTRONOMY and PHYSICAL GEOGRAPHY; comprehending an Account of the Principal Towns, Climate, Soil, Productions, Religion, Education, Government, and Population of the various Countries. With a Compendium of Sacred Geography, Problems on the Globes, Exercises, etc. *Carefully Revised.* 2s. 6d.; or with Four Maps, 2s. 9d.

## An Introductory Geography, for Junior Pupils. By Dr

JAMES DOUGLAS, lately Teacher of English, Great King Street, Edinburgh. *Carefully Revised.* 6d.

## Dr Douglas's Progressive Geography. 144 pages, 1s.

*In the Press.*

## Dr Douglas's Text-Book of Geography, containing the

PHYSICAL and POLITICAL GEOGRAPHY of all the Countries of the Globe. Systematically arranged. 2s. 6d.; or with ten Coloured Maps, 3s. *Carefully Revised.*

## First Book of Geography; being an Abridgment of
Dr Reid's Rudiments of Modern Geography; with an Outline of the
Geography of Palestine. *Carefully Revised.* 6d.

This work has been prepared for the use of young pupils. It is a suitable
and useful companion to Dr Reid's Introductory Atlas.

## Rudiments of Modern Geography. By ALEX. REID,
LL.D., late Head Master of the Edinburgh Institution. With
Plates, Map of the World. *Carefully Revised.* 1s.; or with Five
Maps, 1s. 3d. *Now printed from a larger type.*

The names of places are accented, and they are accompanied with short
descriptions, and occasionally with the mention of some remarkable event.
To the several countries are appended notices of their physical geography,
productions, government, and religion. The Appendix contains an outline of
ancient geography, an outline of sacred geography, problems on the use of
the globes, and directions for the construction of maps.

## Dr Reid's Outline of Sacred Geography. 6d.

This little work is a manual of Scripture Geography for young persons.
It is designed to communicate such a knowledge of the places mentioned in
holy writ as will enable children more clearly to understand the sacred narrative. It contains references to the passages of Scripture in which the
most remarkable places are mentioned, notes chiefly historical and descriptive, and a Map of the Holy Land in provinces and tribes.

## Murphy's Bible Atlas of 24 MAPS, with Historical
Descriptions. 1s. 6d. coloured.

*Witness.*—" We recommend this Atlas to teachers, parents, and individual
Christians, as a comprehensive and cheap auxiliary to the intelligent reading
of the Scriptures.

## Ewing's System of Geography. *Carefully Revised.* 4s. 6d.;
with 14 Maps, 6s.

Besides a complete treatise on the science of geography, this work contains
the elements of astronomy and of physical geography, and a variety of problems to be solved by the terrestrial and celestial globes. At the end is a
pronouncing Vocabulary, in the form of a gazetteer, containing the names
of all the places in the work.

## Elements of Astronomy: adapted for Private Instruction
and Use of Schools. By HUGO REID, Member of the College of
Preceptors. With 65 Wood Engravings. 3s.

## Reid's Elements of Physical Geography; with Outlines
of GEOLOGY, MATHEMATICAL GEOGRAPHY, and ASTRONOMY, and
Questions for Examination. With numerous Illustrations, and a
large coloured Physical Chart of the Globe. 1s.

REVISED EDITIONS OF SCHOOL ATLASES.

## A General Atlas of Modern Geography; 29 Maps,
Coloured. By THOMAS EWING. 7s. 6d.

## School Atlas of Modern Geography. Maps 4to, folded
8vo, Coloured. By John WHITE, F.E.I.S., Author of "Abstract of General Geography," etc. 6s.

## White's Elementary Atlas of Modern Geography.
4to, 10 Maps, Coloured. 2s. 6d.

CONTENTS.—1. The World; 2. Europe; 3. Asia; 4. Africa; 5. North America; 6. South America; 7. England; 8. Scotland; 9. Ireland; 10. Palestine.

## A School Atlas of Modern Geography. 4to, 16 Maps,
Coloured. By ALEXANDER REID, LL.D., late Head Master of the Edinburgh Institution, etc. 5s.

## Reid's Introductory Atlas of Modern Geography.
4to, 10 Maps, Coloured. 2s. 6d.

CONTENTS.—1. The World; 2. Europe; 3. Asia; 4. Africa; 5. North America; 6. South America; 7. England; 8. Scotland; 9. Ireland; 10. Palestine.

# HISTORY.

THE works in this department have been prepared with the greatest care. They will be found to include Class-books for Junior and Senior Classes in all the branches of History generally taught in the best schools. While the utmost attention has been paid to accuracy, the narratives have in every case been rendered as instructive and pleasing as possible, so as to relieve the study from the tediousness of a mere dry detail of facts.

## A Concise History of England in Epochs. By J. F.
CORKRAN. With Maps and Genealogical and Chronological Tables, and comprehensive Questions to each Chapter. 2s. 6d.

*₊* *Intended chiefly for the Senior Classes of Schools, and for the Junior Students of Training Colleges.*

In this History of England the writer has endeavoured to convey a broad and full impression of its great Epochs, and to develop with care, but in subordination to the rest of the narrative, the growth of Law and of the Constitution. He has summarized events of minor importance; but where illustrious characters were to be brought into relief, or where the story of some great achievement merited a full narration, he has occupied more space than the length of the history might seem to justify; for it is his belief that a mere narration of the *Deeds* of England in her struggles for liberty and for a high place among the nations of the world, is more fertile in instruction to youth, and more stimulating to a healthy and laudable ambition than any other mode of treating our past. . . . .
Recent events have been treated with more than usual fulness.

## History of England for Junior Classes; with Questions

for Examination. Edited by HENRY WHITE, B.A., Trinity College, Cambridge, M.A. and Ph. Dr. Heidelberg. 1s. 6d.

*Athenæum.*—" A cheap and excellent history of England, admirably adapted for the use of junior classes. Within the compass of about a hundred and eighty duodecimo pages, the editor has managed to give all the leading facts of our history, dwelling with due emphasis on those turning points which mark our progress both at home and abroad. The various changes that have taken place in our constitution are briefly but clearly described. It is surprising how successfully the editor has not merely avoided the obscurity which generally accompanies brevity, but invested his narrative with an interest too often wanting in larger historical works. The information conveyed is thoroughly sound; and the utility of the book is much increased by the addition of examination questions at the end of each chapter. Whether regarded as an interesting reading-book or as an instructive class-book, this history deserves to rank high. When we add, that it appears in the form of a neat little volume at the moderate price of eighteenpence, no further recommendation will be necessary."

## History of Great Britain and Ireland; with an Account

of the Present State and Resources of the United Kingdom and its Colonies. With Questions for Examination, and a Map. By Dr WHITE. 3s.

*Athenæum.*—" A carefully compiled history for the use of schools. The writer has consulted the more recent authorities: his opinions are liberal, and on the whole just and impartial: the succession of events is developed with clearness, and with more of that picturesque effect which so delights the young than is common in historical abstracts. The book is accompanied by a good map. For schools, parish and prison libraries, workmen's halls, and such institutions, it is better adapted than any abridgment of the kind we know."

## History of Scotland for Junior Classes; with Questions

for Examination. Edited by Dr WHITE. 1s. 6d.

## History of Scotland, from the Earliest Period to the Present

Time. With Questions for Examination. Edited by Dr WHITE. 3s. 6d.

## History of France; with Questions for Examination, and a

Map. Edited by Dr WHITE. 3s. 6d.

*Athenæum.*—" We have already had occasion to speak favourably of Dr White's ' History of Great Britain and Ireland.' The perusal of the present work has given us still greater pleasure. . . . Dr White is remarkably happy in combining convenient brevity with sufficiency of information, clearness of exposition, and interest of detail. He shows great judgment in apportioning to each subject its due amount of consideration."

## Outlines of Universal History. Edited by Dr

WHITE. 2s.

*Spectator.*—" Distinct in its arrangement, skilful in its selection of leading features, close and clear in its narrative."

## Dr White's Elements of Universal History, on a New
and Systematic Plan. In THREE PARTS. PART I. Ancient History;
Part II. History of the Middle Ages; Part III., Modern History.
With a Map of the World. 7s.; or in Parts, 2s. 6d. each.

This work contains numerous synoptical and other tables, to guide the
researches of the student, with sketches of literature, antiquities, and manners
during each of the great chronological epochs.

## Outlines of the History of Rome; with Questions for
Examination. Edited by Dr WHITE. 1s. 6d.

*London Review.*—"This abridgment is admirably adapted for the use of
schools,—the best book that a teacher could place in the hand of a youthful
student."

## Sacred History, from the Creation of the World to the
Destruction of Jerusalem. With Questions for Examination.
Edited by Dr WHITE. 1s. 6d.

*Baptist Magazine.*—"An interesting epitome of sacred history, calculated to
inspire the young with a love of the divine records, as well as to store the
mind with knowledge."

## Elements of General History, Ancient and Modern. To
which are added, a Comparative View of Ancient and Modern
Geography, and a Table of Chronology. By ALEXANDER FRASER
TYTLER, Lord Woodhouselee, formerly Professor of History in the
University of Edinburgh. *New Edition, with the History continued.*
With two large Maps, etc. 3s. 6d.

## Watts' Catechism of Scripture History, and of the
Condition of the Jews from the Close of the Old Testament to
the Time of Christ. With INTRODUCTION by W. K. TWEEDIE,
D.D. 2s.

## Simpson's History of Scotland; with an Outline of the
British Constitution, and Questions for Examination at the end of
each Section. 3s. 6d.

## Simpson's Goldsmith's History of England; with the
Narrative brought down to the Middle of the Nineteenth Century.
To which is added an Outline of the British Constitution. With
Questions for Examination at the end of each Section. 3s. 6d.

## Simpson's Goldsmith's History of Greece. With
Questions for Examination at the end of each Section. 3s. 6d.

## Simpson's Goldsmith's History of Rome. With Questions
for Examination at the end of each Section. 3s. 6d.

## WRITING, ARITHMETIC, AND BOOK-KEEPING.

THIS section will be found to contain works in extensive use in many of the best schools in the United Kingdom. The successive editions have been carefully revised and amended.

### Practical Arithmetic for Junior Classes. By HENRY
G. C. SMITH, Teacher of Arithmetic and Mathematics in George Heriot's Hospital. 64 pages, 6d. stiff wrapper. *Answers*, 6d.

*From the Rev.* PHILIP KELLAND, *A.M., F.R.SS. L. & E., late Fellow of Queens' College, Cambridge, Professor of Mathematics in the University of Edinburgh.*

"I am glad to learn that Mr Smith's Manual for Junior Classes, the MS. of which I have examined, is nearly ready for publication. Trusting that the Illustrative Processes which he has exhibited may prove as efficient in other hands as they have proved in his own, I have great pleasure in recommending the work, being satisfied that a better Arithmetician and a more judicious Teacher than Mr Smith is not to be found."

### Practical Arithmetic for Senior Classes; being a Continuation of the above. By HENRY G. C. SMITH. 2s. *Answers*, 6d. KEY, 2s. 6d.

\*\*\* *The Exercises in both works, which are copious and original, have been constructed so as to combine interest with utility. They are accompanied by illustrative processes.*

*English Journal of Education.*—"There are, it must be confessed, few. good books on arithmetic, but this certainly appears to us to be one of them. It is evidently the production of a practical man, who desires to give his pupils a thorough knowledge of his subject. The Rules are laid down with much precision and simplicity, and the illustrations cannot fail to make them intelligible to boys of ordinary capacity."

### Lessons in Arithmetic for Junior Classes. By JAMES
TROTTER. 66 pages, 6d. stiff wrapper; or 8d. cloth. *Answers*, 6d.

This book was *carefully revised*, and *enlarged* by the introduction of Simple Examples of the various rules, worked out at length, and fully explained, and of Practical Exercises, by the Author's son, Mr Alexander Trotter, Teacher of Mathematics, etc., Edinburgh; and to the present edition Exercises on the proposed Decimal Coinage have been added.

### Lessons in Arithmetic for Advanced Classes; being
a Continuation of the Lessons in Arithmetic for Junior Classes. Containing Vulgar and Decimal Fractions; Simple and Compound Proportion, with their Applications; Simple and Compound Interest; Involution and Evolution, etc. By ALEXANDER TROTTER. New Edition, with Exercises on the proposed Decimal Coinage. 76 pages, 6d. in stiff wrapper; or 8d. cloth. *Answers*, 6d.

Each subject is also accompanied by an example fully worked out and minutely explained. The Exercises are numerous and practical.

## A Complete System of Arithmetic, Theoretical and
Practical; containing the Fundamental Rules, and their Application
to Mercantile Computations; Vulgar and Decimal Fractions; Involution and Evolution; Series; Annuities, Certain and Contingent.
By Mr TROTTER. 3s. KEY, 4s. 6d.

\*\*\* *All the 3400 Exercises in this work are new. They are applicable to the
business of real life, and are framed in such a way as to lead the pupil to reason
on the matter. There are upwards of 200 Examples wrought out at length and
minutely explained.*

## Ingram's Principles of Arithmetic, and their Application
to Business explained in a Popular Manner, and clearly Illustrated
by Simple Rules and Numerous Examples. *Remodelled and greatly
Enlarged*, with Exercises on the proposed Decimal Coinage. By
ALEXANDER TROTTER, Teacher of Mathematics, etc., Edinburgh. 1s.
KEY, 2s.

*Each rule is followed by an example wrought out at length, and is illustrated
by a great variety of practical questions applicable to business.*

## Melrose's Concise System of Practical Arithmetic;
containing the Fundamental Rules and their Application to Mercantile Calculations; Vulgar and Decimal Fractions; Exchanges;
Involution and Evolution; Progressions; Annuities, Certain and
Contingent, etc. *Re-arranged, Improved, and Enlarged*, with Exercises on the proposed Decimal Coinage. By ALEXANDER TROTTER,
Teacher of Mathematics, etc., in Edinburgh. 1s. 6d. KEY, 2s. 6d.

*Each Rule is followed by an example worked out at length, and minutely
explained, and by numerous practical Exercises.*

## Hutton's Arithmetic and Book-keeping. 2s. 6d.

## Hutton's Book-keeping, by TROTTER. 2s.

*Sets of Ruled Writing Books,*—Single Entry, per set, 1s. 6d.; Double Entry,
per set, 1s. 6d.

## Stewart's First Lessons in Arithmetic, for Junior Classes;
containing Exercises in Simple and Compound Quantities arranged
so as to enable the Pupil to perform the Operations with the greatest
facility and correctness. With Exercises on the Proposed Decimal
Coinage. 6d. stiff wrapper. *Answers*, 6d.

## Stewart's Practical Treatise on Arithmetic, *Arranged
for Pupils in Classes.* With Exercises on the proposed Decimal
Coinage. 1s. 6d. This work includes the Answers; with Questions
for Examination. KEY, 2s.

## Gray's Introduction to Arithmetic; with Exercises on
the proposed Decimal Coinage. 10d. bound in leather. KEY, 2s.

**Lessons in Arithmetic for Junior Classes.** By JAMES
MACLAREN, Master of the Classical and Mercantile Academy,
Hamilton Place, Edinburgh. 6d. stiff wrapper.
The Answers are annexed to the several Exercises.

**Maclaren's Improved System of Practical Book-**
KEEPING, arranged according to Single Entry, and adapted to
General Business. Exemplified in one set of Books. 1s. 6d.
*A Set of Ruled Writing Books, expressly adapted for this work, 1s. 6d.*

**Scott's First Lessons in Arithmetic.** 6d. stiff wrapper.
*Answers,* 6d.

**Scott's Mental Calculation Text-book.** Pupil's Copy, 6d.
Teacher's Copy, 6d.

**Copy Books,** in a Progressive Series. By R. SCOTT, late
Writing-Master, Edinburgh. Each containing 24 pages. Price:
Medium paper, 3d; Post paper, 4d.

**Scott's Copy Lines,** in a Progressive Series, 4d. each.

**The Principles of Gaelic Grammar;** with the Definitions,
Rules, and Examples, clearly expressed in English and Gaelic;
containing copious Exercises for Reading the Language, and for
Parsing and Correction. By the Rev. JOHN FORBES, late Minister
of Sleat. 3s. 6d.

## MATHEMATICS, NATURAL PHILOSOPHY, ETC.

**Ingram's Concise System of Mathematics,** Theoretical
and Practical, for Schools and Private Students. Improved by
JAMES TROTTER. With 340 Woodcuts. 4s. 6d. KEY, 3s 6d.

**Trotter's Manual of Logarithms and Practical Mathe-**
MATICS, for Students, Engineers, Navigators, and Surveyors. 3s.

**A Complete System of Mensuration;** for Schools, Private
Students, and Practical Men. By ALEX. INGRAM. Improved by
JAMES TROTTER. 2s.

**Ingram and Trotter's Euclid.** 1s. 6d.

**Ingram and Trotter's Elements of Algebra,** Theoretical
and Practical, for Schools and Private Students. 3s.

## Introductory Book of the Sciences. By JAMES NICOL,
F.R.S.E., F.G.S., Professor of Natural History in the University of Aberdeen. With 106 Woodcuts. 1s. 6d.

### SCHOOL SONGS WITH MUSIC,
By T. M. HUNTER, Director to the Association for the Revival of Sacred Music in Scotland.

## Elements of Vocal Music : An Introduction to the Art of Reading Music at Sight. Price 6d.

\*₊\* *This Work has been prepared with great care, and is the result of long practical experience in teaching. It is adapted to all ages and classes, and will be found considerably to lighten the labour of both teacher and pupil. The exercises are printed in the standard notation, and the notes are named as in the original Sol-fa System.*

CONTENTS.—Music Scales.—Exercises in Time.—Syncopation.—The Chromatic Scale.—Transposition of Scale.—The Minor Scale.—Part Singing.—Explanation of Musical Terms.

## Hunter's School Songs. With Preface by Rev. JAMES
CURRIE, Training College, Edinburgh.

FOR JUNIOR CLASSES : 60 Songs, principally set for two voices. 4d.—*Second Series:* 63 Songs. 4d.

FOR ADVANCED CLASSES : 44 Songs, principally set for three voices. 6d.—*Second Series:* 46 Songs. 6d.

## School Psalmody ; containing 58 Pieces arranged for three voices. 4d.

### GEOMETRICAL DRAWING.

## The First Grade Practical Geometry. Intended chiefly for the use of Drawing Classes in Elementary Schools taught in connexion with the Department of Science and Art. By JOHN KENNEDY, Head Master of Dundee School of Art. 6d.

## School Register. PUPIL'S DAILY REGISTER OF MARKS.
*Improved Edition.* Containing Spaces for 48 Weeks ; to which are added, Spaces for a Summary and Order of Merit for each Month, for each Quarter, and for the Year. For Schools in general, and constructed to furnish information required by Government. 2d.

## School Register of Attendance, Absence, and Fees :
adapted to the Provisions of the Revised Code, by MORRIS F. MYRON. Each folio will serve 50 pupils for a Quarter. 1s.

---

---

CLASS-BOOKS BY CHAS. HENRI SCHNEIDER, F.E.I.S.,
M.C.P.,

Senior French Master in the Edinburgh High School, the Merchant Company's Educational Institution for Young Ladies, the School of Arts and Watt Institution, etc.; French Examiner to the Educational Institute of Scotland, etc.

### Schneider's First Year's French Course. 1s. 6d.

*₊* This work forms a Complete Course of French for Beginners, and comprehends Grammatical Exercises, with Rules; Reading Lessons, with Notes; Dictation; Exercises in Conversation; and a Vocabulary of all the Words in the Book.

### The Edinburgh High School French Conversation-

GRAMMAR, arranged on an entirely New Plan, with Questions and Answers. *Dedicated, by permission, to Professor Max Müller.* 3s. 6d. KEY, 2s. 6d.

### The Edinburgh High School New Practical French

READER: Being a Collection of Pieces from the best French Authors. With Questions and Notes, enabling both Master and Pupil to converse in French. 3s. 6d.

### The Edinburgh High School French Manual of

CONVERSATION and COMMERCIAL CORRESPONDENCE. 2s. 6d.

In this work, Phrases and Idiomatic Expressions which are used most frequently in the intercourse of every-day life have been carefully collected. Care has been taken to avoid what is trivial and obsolete, and to introduce all the modern terms relative to railways, steamboats, and travelling in general.

### Écrin Littéraire: Being a Collection of LIVELY ANEC-

DOTES, JEUX DE MOTS, ENIGMAS, CHARADES, POETRY, etc., to serve as Readings, Dictation, and Recitation. 3s. 6d.

*Letter from* PROFESSOR MAX MÜLLER, *University of Oxford, May* 1867.

"MY DEAR SIR,—I am very happy to find that my anticipations as to the success of your Grammar have been fully realized. Your book does not require any longer a godfather; but if you wish me to act as such, I shall be most happy to have my name connected with your prosperous child.—Yours very truly, MAX MÜLLER.

" To Mons. C. H. Schneider, Edinburgh High School."

---

### The French New Testament. The most approved

PROTESTANT VERSION, and the one in general use in the FRENCH REFORMED CHURCHES. Pocket Edition, roan, gilt edges, 1s. 6d.

### Chambaud's Fables Choisies. With a Vocabulary

containing the meaning of all the Words. By SCOT and WELLS. 2s.

### Le Petit Fablier. With Vocabulary. For Junior Classes.

By G. M. GIBSON, late Rector of the Bathgate Academy. 1s. 6d.

## Standard Pronouncing Dictionary of the French and
ENGLISH LANGUAGES. In Two PARTS. Part I. *French and English.*— Part II. *English and French.* By GABRIEL SURENNE, late Professor in the Scottish Naval and Military Academy, etc. The First Part comprehends Words in ·Common Use, Terms connected with Science and the Fine Arts, Historical, Geographical, and Biographical Names, with the Pronunciation according to the French Academy and the most eminent Lexicographers and Grammarians. The Second Part is an ample Dictionary of English words, with the Pronunciation according to the best Authorities. The whole is preceded by a Practical and Comprehensive System of French Pronunciation. 7s. 6d., strongly bound.

*The Pronunciation is shown by a different spelling of the Words.*

## Surenne's French-English and English-French
DICTIONARY, without the Pronunciation. 3s. 6d. strongly bound.

## Surenne's Fenelon's Telemaque. 2 vols, 1s. each, stiff
wrapper; or bound together, 2s. 6d.

## Surenne's Voltaire's Histoire de Charles XII.
1s. stiff wrapper; or 1s. 6d. bound.

## Surenne's Voltaire's Histoire de Russie sous Pierre
LE GRAND. 2 vols, 1s. each, stiff wrapper; or bound together, 2s. 6d.

## Surenne's Voltaire's la Henriade. 1s. stiff wrapper;
or 1s. 6d. bound.

## Surenne's New French Dialogues; With an Introduction to French Pronunciation, a Copious Vocabulary, and Models of Epistolary Correspondence. *Pronunciation marked throughout.* 2s.

## Surenne's New French Manual and Traveller's
COMPANION. Containing an Introduction to French Pronunciation; a Copious Vocabulary; a very complete Series of Dialogues on Topics of Every-day Life; Dialogues on the Principal Continental Tours, and on the Objects of Interest in Paris; with Models of Epistolary Correspondence. Intended as a Class-book for the Student and a Guide to the Tourist. Map. *Pronunciation marked throughout.* 3s. 6d.

## Surenne's Pronouncing French Primer. Containing
the Principles of French Pronunciation, a Vocabulary of easy and familiar Words, and a selection of Phrases. 1s. 6d. stiff wrapper.

## Surenne's Moliere's l'Avare: Comédie. 1s. stiff wrapper; or 1s. 6d. bound.

## Surenne's. Moliere's le Bourgeois Gentilhomme:
Comédie. 1s. stiff wrapper; or 1s. 6d. bound.

**Surenne's Moliere's Le Misanthrope:** Comédie. **Le**
MARIAGE FORCE: Comédie. 1s. stiff wrapper; or 1s. 6d. bound.

**Surenne's French Reading Instructor,** *Reduced to* 2s. 6d.

**Hallard's French Grammar.** 3s. 6d. KEY, 3s. 6d.

**Grammar of the French Language.** BY AUGUSTE
BELJAME, B.A., LL.B., Vice-Principal of the Paris International
College. 2s.

**Beljame's Four Hundred Practical Exercises.** Being
a Sequel to Beljame's French Grammar. 2s.

**.*** *Both Books bound together,* 3s. 6d.

The whole work has been composed with a view to conversation, a great
number of the Exercises being in the form of questions and answers.

**First French Class-book,** or a Practical and Easy Method
of learning the FRENCH LANGUAGE, consisting of a series of FRENCH
and ENGLISH EXERCISES, progressively and grammatically arranged.
By JULES CARON, F.E.I.S., French Teacher, Edin. 1s. KEY, 1s.

This work follows the natural mode in which a child learns to speak its own
language, by repeating the same words and phrases in a great variety of forms
until the pupil becomes familiar with their use.

**Caron's First French Reading-book:** Being Easy and
Interesting Lessons, progressively arranged. With a copious Vocabulary of the Words and Idioms in the text. 1s.

**Caron's Principles of French Grammar.** With numerous
Exercises. 2s. KEY, 2s.

*Spectator.*—" May be recommended for clearness of exposition, gradual progression, and a distinct exhibition to the mind through the eye by means of typo-
graphical display: the last an important point where the subject admits of it."

**An Easy Grammar of the French Language.** With
EXERCISES AND DIALOGUES. By JOHN CHRISTISON, Teacher of
Modern Languages. 1s. 4d. KEY, 8d.

**Christison's Recueil de Fables et Contes Choisis,**
à l'Usage de la Jéunesse. 1s. 4d.

**Christison's Fleury's Histoire de France,** Racontée
à la Jeunesse. With Translations of the difficult Passages. 2s. 6d.

**French Extracts for Beginners.** With a Vocabulary
and an Introduction. By F. A. WOLSKI, Master of the Foreign
Language Department in the High School of Glasgow. 2s. 6d.

**Wolski's New French Grammar.** With Exercises. 3s. 6d.

## EDINBURGH ACADEMY CLASS-BOOKS.

THE acknowledged merit of these school-books, and the high reputation of the seminary from which they emanate, almost supersede the necessity of any recommendation. The "Latin" and "Greek Rudiments" form an introduction to these languages at once simple, perspicuous, and comprehensive. The "Latin Rudiments" contain an *Appendix*, which renders the use of a separate work on Grammar quite unnecessary; and the *list of anomalous verbs* in the "Greek Rudiments" is believed to be more extensive and complete than any that has yet appeared in School Grammars of the language. In the "Latin Delectus" and "Greek Extracts" the sentences have been arranged strictly on the *progressive principle*, increasing in difficulty with the advancement of the Pupil's knowledge; while the *Vocabularies* contain an explanation not only of every *word*, but also of every *difficult expression* which is found in the works,—thus rendering the acquisition of the Latin and Greek languages both easy and agreeable. The Selections from Cicero embrace the portions of his works which are best adapted for Scholastic tuition.

### 1. Rudiments of the Latin Language. 2s.

*∗* *This work forms an introduction to the language, at once simple, perspicuous, and comprehensive.*

### 2. Latin Delectus ; with a Vocabulary containing an Explanation of every Word and Difficult Expression which occurs in the Text. 3s. 6d.

### 3. Rudiments of the Greek Language. 3s. 6d.

### 4. Greek Extracts ; with a Vocabulary containing an Explanation of every Word and of the more Difficult Passages in the Text. 3s. 6d.

### 5. Selections from Cicero. 3s.

### 6. Selecta e Poetis Latinis ; including Extracts from Plautus, Terence, Lucretius, Catullus, Persius, Lucan, Martial, Juvenal, etc. 3s.

### Greek Syntax ; with a Rationale of the Constructions, by JAS. CLYDE, LL.D., one of the Classical Masters of the Edinburgh Academy. With Prefatory Notice by JOHN S. BLACKIE, Professor of Greek in the University of Edinburgh. *4th Edition*, entirely re-written, and enlarged by a Summary for the use of Learners and a chapter on Accents. 4s. 6d.

### Greek Grammar for the Use of Colleges and Schools. By Professor GEDDES, University of Aberdeen. 4s.

The author has endeavoured to combine the clearness and conciseness of the older Greek Grammars with the accuracy and fulness of more recent ones.

### DR HUNTER'S CLASSICS.

1. **Hunter's Ruddiman's Rudiments.** 1s. 6d.
2. **Hunter's Sallust;** with Footnotes and Translations. 1s. 6d.
3. **Hunter's Virgil;** with Notes and other Illustrations. 2s. 6d.
4. **Hunter's Horace.** 2s.
5. **Hunter's Livy.** Books XXI. to XXV. With Critical and Explanatory Notes. *Reduced to 3s.*

**Latin Prose Composition:** The Construction of Clauses, with Illustrations from Cicero and Cæsar; a Vocabulary containing an Explanation of every Word in the Text; and an Index Verborum. By JOHN MASSIE, A.M. 3s. 6d.

**Dymock's Cæsar;** with illustrative Notes, a Historical and Geographical Index, and a Map of Ancient Gaul. 4s.

**Dymock's Sallust;** with Explanatory Footnotes and a Historical and Geographical Index. 2s.

**Cæsar;** with Vocabulary explaining every Word in the Text, Notes, Map, and Historical Memoir. By WILLIAM M'DOWALL, late Inspector of the Heriot Foundation Schools, Edinburgh. 3s.

**M'Dowall's Virgil;** with Memoir, Notes, and Vocabulary explaining every Word in the Text. 3s.

**Neilson's Eutropius et Aurelius Victor;** with Vocabulary containing the meaning of every Word that occurs in the Text. *Revised* by WM. M'DOWALL. 2s.

**Lectiones Selectae:** or, Select Latin Lessons in Morality, History, and Biography: for the use of Beginners. With a Vocabulary explaining every Word in the Text. By C. MELVILLE, late of the Grammar School, Kirkcaldy. 1s. 6d.

**Macgowan's Lessons in Latin Reading.** In TWO PARTS. Part I., *Improved* by H. FRASER HALLE, LL.D. 2s. 17th Edition. Part II. 2s. 6d. The Two Courses furnish a complete Latin Library of Reading, Grammar, and Composition for Beginners, consisting of Lessons which advance in difficulty by easy gradations, accompanied by Exercises in English to be turned into Latin. Each volume contains a complete Dictionary adapted to itself.

# Latin and Greek. 23

## Mair's Introduction to Latin Syntax: with Illustrations
by Rev. ALEX. STEWART, LL.D.; an English and Latin Vocabulary, for the assistance of the Pupil in translating into Latin the English Exercises on each Rule; and an Explanatory Vocabulary of Proper Names. 3s.

## Stewart's Cornelius Nepos; with Notes, Chronological
Tables, and a Vocabulary explaining every Word in the Text. 3s.

## Ainsworth's Latin Dictionary. Edited by WM. DUNCAN,
E.C.P. 1070 pages. 9s. strongly bound.

This edition contains a copious index of proper names, a complete list of Latin abbreviations, and other important and useful tables.

## Duncan's Greek Testament. 3s. 6d.

## Beza's Latin Testament. Revised by the late ADAM
DICKINSON, A.M. 3s. 6d.

## Xenophon's Anabasis, Books I. and II.; with Vocabulary
giving an Explanation of every Word in the Text, and a Translation of the more difficult Phrases. By JAMES FERGUSSON, M.D., late Rector of the West End Academy, Aberdeen. 2s. 6d.

*Athenæum.*—"The text of this admirable little work is that of Dindorf, and the punctuation generally that of Poppo. Its principal excellence as an introduction to the study of Greek consists in the copious, correct, and well arranged Vocabulary at the end. This contains good translations of difficult passages, with exact information upon points of antiquities derived from the best and most modern authorities."

## Grammatical Exercises on the Moods, Tenses, and
SYNTAX OF ATTIC GREEK. With a Vocabulary containing the meaning of every Word in the Text. On the plan of Professor Ferguson's Latin "Grammatical Exercises." By Dr FERGUSSON. 3s. 6d. KEY, 3s. 6d.

*₊* *This work is intended to follow the Greek Rudiments.*

## Homer's Iliad—Greek, from Bekker's Text. Edited
by the Rev. W. VEITCH, Author of "Greek Verbs, Irregular and Defective," etc. 3s. 6d.

## Homer's Iliad, Books I., VI., XX., and XXIV.; with
Vocabulary giving an Explanation of every Word in the Text, and a Translation of the more difficult Passages. By Dr FERGUSSON. 3s. 6d.

**24**                *Latin and Greek.*

### LATIN ELEMENTARY WORKS AND CLASSICS.

Edited by GEORGE FERGUSON, LL.D., lately Professor of Humanity in King's
College and University of Aberdeen, and formerly one of the
Masters of the Edinburgh Academy.

1. **Ferguson's Grammatical Exercises.** With Notes,
and a Vocabulary explaining every Word in the Text. 2s. KEY, 2s.

2. **Ferguson's Introductory Latin Delectus**; Intended
to follow the Latin Rudiments; with a Vocabulary containing an
Explanation of every Word and of every Difficult Expression. 2s.

3. **Ferguson's Ovid's Metamorphoses.** With Explanatory
Notes and an Index, containing Mythological, Geographical, and
Historical Illustrations. 2s. 6d.

4. **Ferguson's Ciceronis Orationes Selectae.** Containing
pro Lege Manilia, IV. in Catilinam, pro A. L. Archia, pro T. A.
Milone. Ex Orellii recensione. 1s. 6d.                    .

5. **Ferguson's Ciceronis Cato Major** sive de Senectute,
Laelius sive de Amicitia, Somnium Scipionis, et Epistolae Selectae.
Ex Orellii recensione. 1s. 6d.

6. **Ferguson's Ciceronis de Officiis.** Ex Orellii recensione. 1s. 6d.

**The Port-Royal Logic.** Translated from the French,
with Introduction, Notes, and Appendix. By THOMAS SPENCER
BAYNES, B.A., Professor of Logic, Rhetoric, and Metaphysics,
United College of St Salvator and St Leonard, St Andrews. 4s.

# ITALIAN.

**Theoretical and Practical Italian Grammar;** with
Numerous Exercises and Examples, illustrative of every Rule, and
a Selection of Phrases and Dialogues. By E. LEMMI, LL.D., Italian
Tutor to H. R. H. the Prince of Wales. 5s.—KEY, 5s.

*From* COUNT SAFFI, *Professor of the Italian Language at Oxford.*—"I have
adopted my Grammar for the elementary instruction of students of Italian
in the Taylor Institution, and find it admirably adapted to the purpose, as
well for the order and clearness of the rules, as for the practical excellence
and ability of the exercises with which you have enriched it."

**PUBLISHED BY OLIVER AND BOYD, EDINBURGH;**

SOLD ALSO BY SIMPKIN, MARSHALL, AND CO., LONDON, AND ALL BOOKSELLERS.